Lecture Notes in Mathematics

Edited by A. Dold and B. Eckmann

655

Ricardo Baeza

Quadratic Forms Over Semilocal Rings

Springer-Verlag
Berlin Heidelberg New York 1978

Author
Ricardo Baeza
Mathematisches Institut FB 9
Universität des Saarlandes
D-6600 Saarbrücken

AMS Subject Classifications (1970): primary: 10 C 05, 10 E 04, 10 E 08

ISBN 3-540-08845-8 Springer-Verlag Berlin Heidelberg New York
ISBN 0-387-08845-8 Springer-Verlag New York Heidelberg Berlin

Printing and binding: Beltz Offsetdruck, Hemsbach/Bergstr.
2141/3140-543210

Preface

The algebraic theory of quadratic forms originates from the well-known paper [W] of Witt (1937), where he introduced for the first time the so called Witt ring of a field, i.e. the ring of all quadratic forms over a field with respect to a certain equivalence relation (compare §4, chap.I in this work). The study of this ring and related questions is essentially what we understand by the algebraic theory of quadratic forms. Thirty years after the appearance of Witt's work, Pfister succeeded in his important papers $[Pf]_{1,2,3}$ in giving the first results on the structure of the Witt ring of a field of characteristic not 2. Since Pfister's work appeared twelve years ago, a lot researchs on this subject have been made. Lam succeeded in writing down many of these researchs in his fine book [L], which is perhaps today the best source to which a student of the algebraic theory of quadratic forms may turn for a comprehensive treatment of this subject. Besides of the theory over fields a corresponding theory of quadratic forms over more general domains has been growing up. We cite in particular Knebusch's work on the related subject of symmetric bilinear forms (see $[K]_{1,..,7}$).
The present work deals with the algebraic theory of quadratic forms over semi local rings. We have tried to give a treatment which works for any characteristic, i.e. we do not make any assumption about 2. If 2 is not a unit, then in general quadratic forms behave better than bilinear forms, because the former have much more automorphisms (for example an anisotropic bilinear space over a field of characteristic 2 has only one automorphism). This fact has been exploited throughout in this work (see §3,5 in chap.III and §3,4 in chap.IV). Of course our results cannot go so far as in the field case, because over semi local rings we do not have to our disposal one of the most powerful methods of the theory over fields, namely the transcendental method. For example it would be very interesting to have an elementary proof (i.e. without transcendental methods) of the "Hauptsatz" of Arason-Pfister (see[Ar-Pf])or even of Krull-intersection theorem for the Witt ring.
Our treatment is rather self contained. We only suppose the reader to be acquainted with the most elementary facts of the theory of quadratic forms over fields (for example as given in Dieudonne's book $[D]_1$) and with the current results of the theory of Azumaya algebras (see [Ba]).

We have used only standard notation. For example $\mathbb{N}$, $\mathbb{Z}$, $\mathbb{Q}$, $\mathbb{R}$ denote the set of non zero integers, the ring of integers, the field of rationals and the field of real numbers, respectively. Moreover we use the following notation: ch = characteristic, dim = dimension, Ker = kernel, Im = image, $\oplus$ = direct sum, Π = direct product, $\otimes$ = tensor product, etc.

R. Baeza
Mathematisches Institut, FB 9
Universität des Saarlandes, Saarbrücken

Contents

CHAPTER I

Quadratic forms over rings

§ 1. Definitions.

Let A be a commutative ring with 1. Let $\underline{P}(A)$ denote the category of finitely generated projective A-modules with the operations $\oplus$ = direct sum and $\otimes_A$ = tensor product (henceforth we shall use the unadorned tensor product $\otimes$ instead of $\otimes_A$ if the ring A is fixed). For every $M \in \underline{P}(A)$ we call M^* the dual A-module $\mathrm{Hom}_A(M,A) \in \underline{P}(A)$.

(1.1) Definition. Let $M \in \underline{P}(A)$. A bilinear form $b: M \times M \to A$ is called symmetric, if $b(x,y) = b(y,x)$ for all $x,y \in M$. The pair (M,b), consisting of a module $M \in \underline{P}(A)$ and a symmetric bilinear form b on M, will be called a bilinear module over A. We frequently write b instead of (M,b).

Let (M,b) be a bilinear module over A. For every $x \in M$ we define $d_b(x) \in M^*$ by the formula $d_b(x)(y) = b(x,y)$ for all $y \in M$. Thus we obtain a A-linear map $d_b: M \to M^*$. The bilinear module (M,b) is called non singular or simply bilinear space, if d_b is an isomorphism. An isomorphism between two bilinear modules (M_1,b_1) and (M_2,b_2) is given by a linear isomorphism $f: M_1 \xrightarrow{\sim} M_2$ with

$$b_1(x,y) = b_2(f(x),f(y))$$

for all $x,y \in M_1$. Then we write $(M_1,b_1) \cong (M_2,b_2)$.

(1.2) Remark. Let $M = Ae_1 \oplus \cdots \oplus Ae_n$ be a free n-dimensional A-module with a symmetric bilinear form b on it. The form b is determined by the matrix (b_{ij}), where $b_{ij} = b(e_i,e_j)$, because we have

$$b(x,y) = \sum_{i,j=1}^{n} b_{ij} x_i x_j$$

for $x = \sum_{i=1}^{n} x_i e_i$, $y = \sum_{i=1}^{n} y_i e_i$. Conversely, using this formula we obtain from every symmetric $n \times n$-matrix (b_{ij}) over A a symmetric bilinear form on M, and the bilinear module (M,b) is non singular if and only if $\det(b_{ij}) \in A^*$ = group of units of A.

(1.3) Localisation. Let max(A) be the set of all maximal ideals of A. For $m \in \max(A)$ and $M \in \underline{P}(A)$ let M_m denote the localisation of M in m. Correspondingly we write M(m) for the reduction of M modulo m, i.e. $M(m) = M/mM$. Then $M_m \in \underline{P}(A_m)$ and consequently M_m is a free A_m-module of finite dimension (see [Bo]$_1$). The map $r_M : \max(A) \to \mathbb{N}$, given by $r_M(m) = \dim_{A_m}(M_m)$, is called the rank-map (or simply the rank) of M.

M has constant rank $n \in \mathbb{N}$, if $r_M(m) = n$ for all $m \in \max(A)$. Let us now consider a bilinear module (M,b) over A. The bilinear form b induces on M_m a symmetric bilinear form b_m over A_m, $b_m : M_m \times M_m \to A_m$, which is defined by

$$b_m\left(\frac{x}{a},\frac{y}{b}\right) = \frac{b(x,y)}{ab}$$

for $x,y \in M$, $a,b \notin m$. The bilinear module (M_m,b_m) is called the localisation of (M,b) in m. For the induced A_m-linear map $d_{b_m} : M_m \to M_m{}^*$ we get $(d_b)_m = d_{b_m}$, since we may identify $(M^*)_m$ with $M_m{}^*$.

The reduction of (M,b) modulo m is defined analogously, i.e. we have a bilinear $A(m) = A/m$-module $b(m) : M(m) \times M(m) \to A/m$, which is given by $b(m)(\bar{x},\bar{y}) = \overline{b(x,y)}$ for all $\bar{x},\bar{y} \in M(m)$. For the induced A(m)-linear map $d_{b(m)} : M(m) \to M(m)^* = M^*(m)$ we have $d_{b(m)} = d_b(m)$. Now using this remarks and the fact, that a bilinear module over a local ring is non singular if and only if its reduction is non singular, we deduce (see [Bo]$_1$, ch.II, §3)

(1.4) Proposition. Let (M,b) be a bilinear module over A. The following statements are equivalent:

i) (M,b) is non singular

ii) (M_m,b_m) is non singular for all $m \in \max(A)$

iii) (M(m),b(m)) is non singular for all $m \in \max(A)$

Now we introduce quadratic forms

(1.5) Definition. A quadratic form on a module $M \in \underline{P}(A)$ is a map $q : M \to A$ with the following properties

i) $q(\lambda x) = \lambda^2 q(x)$ for all $x \in M$, $\lambda \in A$

ii) $b_q(x,y) = q(x+y) - q(x) - q(y)$ defines on M a bilinear form $b_q : M \times M \to A$.

The pair (M,q) is called a quadratic module over A and (M,b_q) the associated bilinear module. If (M,b_q) is non singular we call (M,q) non singular, or, a quadratic space.

An isomorphism between two quadratic modules (M_1,q_1) and (M_2q_2) is given by a linear isomorphism $f: M_1 \xrightarrow{\sim} M_2$ with the property

$$q_1(x) = q_2(f(x))$$

for all $x \in M_1$. We then write $(M_1,q_1) \cong (M_2,q_2)$. Of course in this case the associated bilinear modules are isomorph, too.

(1.6) Remark. Let us assume $2 \in A^*$. Then we have a one to one correspondence between bilinear modules and quadratic modules over A. By this correspondence the bilinear module (M,b) will be identified with the quadratic module (M,q_b), where $q_b(x) = \frac{1}{2} b(x,x)$ for all $x \in M$, and conversely, the quadratic module (M,q) will correspond to the bilinear module (M,b_q). One sees immediately $q_{b_q} = q$ and $b_{q_b} = b$.

Consider now a free A-module $M = Ae_1 \oplus \cdots \oplus Ae_n$ endowed with a quadratic form q. For every $i \neq j$ we set $a_{ij} = b_q(e_i,e_j)$ and for $i = j$ $a_{ii} = q(e_i)$. We obtain a $n \times n$-matrix $[a_{ij}]$, which is called the value-matrix of (M,q) with respect to the basis $\{e_1,\ldots,e_n\}$. This matrix determines q completely, since for $x = \sum_{i=1}^{n} x_i e_i \in M$

$$q(x) = \sum_{1 \leq i \leq j \leq n} a_{ij} x_i x_j$$

For this reason we shall identify the quadratic form q with his value-matrix $[a_{ij}]$, i.e. we set $q = [a_{ij}]$. For example if $M = Ae \oplus Af$ and $q(e) = a$, $q(f) = b$, $b_q(e,f) = 1$, we have

$$q = \begin{bmatrix} a & 1 \\ 1 & b \end{bmatrix}$$

The value-matrix of (M,b_q) is $\begin{pmatrix} 2a & 1 \\ 1 & 2b \end{pmatrix}$, so that (M,q) is non singular, if and only if $1-4ab \in A^*$. In this case we shall write $[a,b]$ for this quadratic space. If $2 = o$ in A, all quadratic spaces $[a,b]$ have the same associated bilinear space $\begin{pmatrix} o & 1 \\ 1 & o \end{pmatrix}$, i.e. in general two non isomorphic quadratic spaces may have the same associated bilinear space. In any case we have the following

(1.7) <u>Proposition</u>. Let (M,q) be a quadratic module. Then there exists a bilinear form $b_o : M \times M \to A$ with $q(x) = b_o(x,x)$ and $b_q(x,y) = b_o(x,y) + b_o(y,x)$ for all $x,y \in M$.

<u>Proof</u>. First we suppose that M is free, i.e. $M \;\; Ae_1 \oplus \cdots \oplus Ae_n$. Let $a_{ii} = q(e_i)$ and $a_{ij} = b_q(e_i,e_j)$ for $i \neq j$. Then we define $b_o : M \times M \to A$ by $b_o(e_i,e_k) = a_{ik}$ if $i \leq k$ and $b_o(e_i,e_k) = o$ if $i > k$. This bilinear form fulfills the conditions of the proposition. Now we treat the general case. Let $N \in \underline{P}(A)$ be a module with $M \oplus N \cong A^m$ for some m. On A^m we define a quadratic form q' as follows: $q'_{|M} = q$ and $q'_{|N} = o$. Using the above mentioned construction we obtain a bilinear form b_o' associated to q'. Then the form $b_o = b_o'{}_{|M \times M}$ has all required properties.

As in the bilinear case we define for every $m \in \max(A)$ the <u>localisation</u> (M_m, q_m) and the <u>reduction</u> $(M(m), q(m))$ of a given quadratic module (M,q) over A. The analogous result to (1.4) is now

(1.8) <u>Proposition</u>. For every quadratic module (M,q) the following statements are equivalent

i) (M,q) is non singular

ii) (M_m, q_m) is non singular for all $m \in \max(A)$

iii) $(M(m), q(m))$ is non singular for all $m \in \max(A)$.

(1.9) <u>Remark</u>. If A has characteristic 2, then it holds $b_q(x,x) = 2q(x) = o$ for every quadratic form q over A, i.e. b_q is an alternating form. Thus if (M,q) is moreover non singular and free, the dimension of M must be even. More general, the rank of any quadratic space (M,q) over a ring A with $2 \in r = \bigcap_{m \in \max(A)} m$ (Jacobson radical of A) is an even function.

§ 2. Operations with quadratic and bilinear forms.

We shall denote the category of bilinear spaces over A by Bil(A) and the category of quadratic spaces over A by Quad(A). In each case the morphisms in these categories are the isomorphisms. We now define on Bil(A) and Quad(A) several operations.

(2.1) Orthogonal sum. For $(M_i,b_i) \in \text{Bil}(A)$ and respectively $(M_i,q_i) \in \text{Quad}(A)$ $(i=1,2)$ we set

$$(M_1,b_1) \perp (M_2,b_2) = (M_1 \oplus M_2,b)$$

$$(M_2,q_1) \perp (M_2,q_2) = (M_1 \oplus M_2,q)$$

where $b(x_1 \oplus x_2, y_1 \oplus y_2) = b_1(x_1,y_1) + b_2(x_2,y_2)$ and $q(x_1 \oplus x_2) = q_1(x_1) + q_2(x_2)$ for all $x_i,y_i \in M_i$, $i=1,2$. We simply write $b_1 \perp b_2$ and $q_1 \perp q_2$ respectively. Obviously $b_1 \perp b_2 \in \text{Bil}(A)$ and $q_1 \perp q_2 \in \text{Quad}(A)$.

(2.2) Operation of Bil(A) on Quad(A)

Take $(M,b) \in \text{Bil}(A)$ and $(E,q) \in \text{Quad}(A)$. They define a new quadratic space

$$(M,b) \otimes (E,q) = (M \otimes_A E, b \otimes q)$$

where $b \otimes q$ is defined on $M \otimes E$ by $b \otimes q(x \otimes y) = b(x,x)q(y)$ for all $x \in M$, $y \in E$. The associated bilinear space to this quadratic space is the tensor product $(M,b) \otimes (E,b_q)$ as defined below.

(2.3) Tensor product in Bil(A) and Quad(A)

For two bilinear spaces (M_1,b_1), $(M_2,b_2) \in \text{Bil}(A)$ we define

$$(M_1,b_1) \otimes (M_2,b_2) = (M_1 \otimes M_2, b_1 \otimes b_2)$$

where $b_1 \otimes b_2(x_1 \otimes x_2, y_1 \otimes y_2) = b_1(x_1,y_1)b_2(x_2,y_2)$ for all $x_i,y_i \in M_i$ $(i = 1,2)$.

Respectively, for two quadratic spaces (M_1,q_1), (M_2,q_2) there is an isomorphism $(M_1,b_{q_1}) \otimes (M_2,q_2) \cong (M_2,b_{q_2}) \otimes (M_1,q_1)$, which allows us to define, up to isomorphism, the quadratic space

$$(M_1,q_1) \circ (M_2,q_2) = (M_1,b_{q_1}) \otimes (M_2,q_2)$$

For these operations we shall abbreviate $b \otimes q$, $b_1 \otimes b_2$ and $q_1 \circ q_2$ respectively. Evidently associativity, commutativity and distributivity hold for these operations (up to isomorphisms). For example

(2.4) Proposition. For $b \in \underline{\mathrm{Bil}}(A)$, $q_1, q_2 \in \underline{\mathrm{Quad}}(A)$ it holds

$$(b \otimes q_1) \circ q_2 \cong q_1 \circ (b \otimes q_2) \cong b \otimes (q_1 \circ q_2)$$

With the category $\underline{\mathrm{Bil}}(A)$ we associate its corresponding Grothendieck-ring $K_o(\underline{\mathrm{Bil}}(A))$ (see [Ba]). We denote this ring by $\hat{W}(A)$ and call it the Witt-Grothendieck-ring of bilinear spaces over A. Correspondly we construct $\hat{W}_q(A) = K_o(\underline{\mathrm{Quad}}(A))$ and call $\hat{W}_q(A)$ the Witt-Grothendieck-ring of quadratic spaces over A. The ring $\hat{W}(A)$ has a unit element, which is represented by the 1-dimensional bilinear space $<1>$, whereas if $2 \notin A^*$ the ring $\hat{W}_q(A)$ has no unit element. Of course if $2 \in A^*$ we shall identify $\hat{W}_q(A)$ with $\hat{W}(A)$ according to the remark (1.6). In general through the operation (2.3) $\hat{W}_q(A)$ becomes a $\hat{W}(A)$-algebra, which will play an important role in this work. If $[q]$ denotes the isomorphism class of the quadratic space q, then the elements of $\hat{W}_q(A)$ are the formal differences $[q_1]-[q_2]$ of classes $[q_1]$, $[q_2]$, where the equality $[q_1]-[q_2] = [p_1]-[p_2]$ holds if and only if there exists $q \in \underline{\mathrm{Quad}}(A)$ with

$$q_1 \perp p_2 \perp q \cong p_1 \perp q_2 \perp q$$

Respectively we have the same facts in $\hat{W}(A)$.

(2.5) Functoriality of the Witt-Grothendieck-rings.

Let $\alpha : A \to B$ be a ring homomorphism with $\alpha(1) = 1$. Take $(M,b) \in \underline{\mathrm{Bil}}(A)$ and $(E,q) \in \mathrm{Quad}(A)$. Then $M \otimes_A B$, $E \otimes_A B \in \underline{P}(B)$, and we define

$(M \otimes_A B, b_B) \in \underline{\mathrm{Bil}}(B)$, $(E \otimes_A B, q_B) \in \underline{\mathrm{Quad}}(B)$ in the following way

$$b_B(x \otimes \lambda, y \otimes \mu) = \alpha[b(x,y)]\lambda\mu$$

$$q_B(z \otimes \lambda) = \alpha[q(z)]\lambda^2$$

$$b_{q_B}(z \otimes \lambda, w \otimes \mu) = \alpha[b_q(z,w)]\lambda\mu$$

for $x,y \in M$, $z,w \in E$, $\lambda,\mu \in B$. It can easily be seen, that $b_B \in \underline{Bil}(B)$ and $q_B \in \underline{Quad}(B)$. We often shall write $b \otimes B$ for b_B and $q \otimes B$ for q_B. This operation defines two additive and multiplicative functors $\alpha^* : \underline{Bil}(A) \to \underline{Bil}(B)$ and $\alpha^* : \underline{Quad}(A) \to \underline{Quad}(B)$, which induce the following homomorphisms

(2.6) $\qquad \alpha^* : \hat{W}(A) \to \hat{W}(B)$

(2.7) $\qquad \alpha^* : \hat{W}_q(A) \to \hat{W}_q(B)$,

respectively.

(2.8) <u>The transfer</u> (see $[Sch]_1$, $[Sch]_2$, [L]).

Let $i : A \to B$ be a ring extension, so that B is a finitely generated projective A-module (with respect to i). We call B/A a <u>Frobenius extension</u>, if there is a A-linear map $s : B \to A$ (called <u>trace map</u>) with the following property: the symmetric bilinear form $\bar{s} : B \times B \to A$ over A, defined by $\bar{s}(b_1,b_2) = s(b_1b_2)$ for all $b_1,b_2 \in B$, is non singular, i.e. $(B,\bar{s}) \in \underline{Bil}(A)$.
Let us now consider a Frobenius extension B/A with trace map $s:B \to A$. If $M \in \underline{P}(B)$, we shall consider M as an A-module via the homomorphism $i : A \to B$ and we shall denote it by M_A. Since $B \in \underline{P}(A)$, it follows that $M_A \in \underline{P}(A)$. Now we define for every $(M,b) \in \underline{Bil}(B)$ resp. $(M,q) \in \underline{Quad}(B)$ the bilinear module $s_*(M,b) = (M_A, s \circ b)$ resp. the quadratic module $s_*(M,q) = (M_A, s \circ q)$. Then we have

(2.9) <u>Proposition</u>. It holds $s_*(M,b) \in \underline{Bil}(A)$ and $s_*(M,q) \in \underline{Quad}(A)$.

<u>Proof</u>. Because of the evident relation $s_*(M,b_q) = (M_A, b_{s \circ q})$, we only need to prove the first statement of the proposition. Let $d_b : M \to M^* = Hom_B(M,B)$ be the associated isomorphism to (M,b). The trace map $s : B \to A$ induces a homomorphism $s^*: Hom_B(M,B)_A \to Hom_A(M_A,A)$ (over A) by $s^*(f)(x) = s[f(x)]$ for all $f \in M^*$, $x \in M$. Since $d_{s \circ b} = s^* \circ d_b$, it

is sufficient to prove the following

(2.10) Lemma. s^* is an isomorphism.

Proof. Injectivity of s^*. Let $f \in \mathrm{Hom}_B(M,b)$ with $s^*(f) = s \circ f = 0$. Then for every $b \in B$ it holds $\bar{s}(b,f(x)) = s(bf(x)) = s(f(bx)) = 0$ (for all $x \in M$), so that $f(x) = 0$ for all $x \in M$ because $\bar{s}$ is non singular, i.e. $f = 0$.

Surjectivity of s^*. Let $g : M_A \to A$ be A-linear. Then for every $x \in M$ we define $\lambda_x : B \to A$ by $\lambda_x(b) = g(bx)$ for all $b \in B$. Since λ_x is A-linear and $\bar{s}$ is non singular, there exists a unique $b_x \in B$ with $\bar{s}(b,b_x) = s(bb_x) = \lambda_x(b)$ for all $b \in B$. Let us define $f : M \to B$ by $f(x) = b_x$. From the uniqueness of b_x it follows, that f is B-linear. Hence $s \circ f(x) = s[f(x)] = s(b_x) = \lambda_x(1) = g(x)$ for all $x \in M$, i.e. $s^*(f) = g$.

The correspondences $(M,b) \to s_*(M,b)$ and $(M,q) \to s_*(M,q)$ define two additive functors

$s_* : \underline{\mathrm{Bil}}(B) \to \underline{\mathrm{Bil}}(A)$ and $s_* : \underline{\mathrm{Quad}}(B) \to \underline{\mathrm{Quad}}(A)$

respectively, which induce two group homomorphisms

$$(2.11)\qquad \begin{aligned} s_* &: \hat{W}(B) \to \hat{W}(A) \\ s_* &: \hat{W}_q(B) \to \hat{W}_q(A) \end{aligned}$$

They are called transfer homomorphisms.

(2.12) Proposition. (Frobenius reciprocity). Let $i : A \to B$ be a Frobenius extension with the trace map $s : B \to A$.

i) For every $x \in \hat{W}(A)$, $y \in \hat{W}(B)$ (or resp. $x \in \hat{W}_q(A)$, $y \in \hat{W}_q(B)$) it holds

$$s_*(i^*(x)\cdot y) = x\cdot s_*(y)$$

ii) For every $x \in W(A)$, $y \in W_q(B)$ it holds

$$s_*(i^*(x)y) = x\, s_*(y)$$

iii) For every $x \in W(B)$, $y \in W_q(A)$ it holds

$$s_*(xi^*(y)) = s_*(x)y$$

Proof. Let us consider $(E,b) \in \underline{Bil}(A)$ and $(M,b') \in \underline{Bil}(B)$, respectively $(E,q) \in \underline{Quad}(A)$ and $(M,q') \in \underline{Quad}(B)$. The forms $s_*(i^*(b) \otimes b')$, respectively $s_*(i^*(q) \circ q')$, are defined on the module $[(E \underset{A}{\otimes} B) \otimes_B M]_A$. On the other hand the forms $b \otimes s_*(b')$ and resp. $q \circ s_*(q')$ are defined on $E \otimes_A M_A$. Now we have a canonical isomorphism of A-modules

$$[(E \otimes_A B) \otimes M]_A \overset{\sim}{\rightarrow} E \otimes_A M_A$$

defined by $(e \otimes \beta) \otimes m \rightarrow e \otimes \beta m$ for all $e \in E$, $\beta \in B$, $m \in M$, and it is easy to see that this isomorphism is an isomorphism for the above considered bilinear, resp. quadratic forms. This proves (i). The other assertions (ii) and (iii) can be proved similarly

(2.13) Remark. As a consequence of (2.12) one sees easily that $s_*[\hat{W}(B)]$ is an ideal in $\hat{W}(A)$, resp. $s_*[\hat{W}_q(B)]$ is a subalgebra of the $\hat{W}(A)$-algebra $\hat{W}_q(A)$. It is easy to show that the images $s_*[\hat{W}(B)]$ and $s_*[\hat{W}_q(B)]$ are independent of the choise of the trace map s. Thus we shall denote it by $I(B,A)$ and $I_q(B,A)$ respectively. On the other hand we set $K(A,B) = \mathrm{Ker}(i^*)$ in $\hat{W}(A)$ and $K_q(A,B) = \mathrm{Ker}(i^*)$ in $\hat{W}_q(A)$. Hence from the proposition (2.12) it follows $K(A,B) \cdot I(B,A) = o$, $K_q(A,B) \circ I_q(B,A) = o$, $K(A,B) \cdot I_q(B,A) = o$ and $I(B,A) \cdot K_q(A,B) = o$.

§ 3. Subspaces.

Let (E,q) be a quadratic module, resp. (E,b) be a bilinear module over the ring A. For every subset $R \subseteq E$ we set $R^\perp = \{x \in E \mid b_q(x,y) = o$ for all $y \in R\}$, resp. $R^\perp = \{x \in E \mid b(x,y) = o$ for all $y \in R\}$. $R^\perp$ is a submodule of E, and in both cases we have the following

(3.1) Lemma. i) If $S \subseteq R \subseteq E$, then $R^\perp \subseteq S^\perp$.

ii) $R \subseteq R^{\perp\perp}$.

iii) $R^\perp = R^{\perp\perp\perp}$.

We call two subsets $S,R \subseteq E$ orthogonal, if $S \subseteq R^{\perp}$ (or equivalently $R \subseteq S^{\perp}$). We call a subset $S \subseteq E$ totally isotropic, if $S \subseteq S^{\perp}$ and in the quadratic case additionally $q(S) = o$. A submodule $U \subset E$ is called subspace of (E,q), resp. (E,b) if U is a direct sumand of E. If we have a direct sum decomposition $E = U \oplus V$ with $U \subseteq V^{\perp}$, we shall write $E = U \perp V$ and we say that E is the orthogonal sum of U and V. The restriction of the quadratic form q, resp. bilinear form b to the subspace $U \subseteq E$ will be denoted by $(U,q_{|U})$, resp. $(U,b_{|U})$. It is clear that if $E = U \perp V$, then $(E,q) \cong (U,q_{|U}) \perp (V,q_{|V})$ respectively $(E,b) \cong (U,b_{|U}) \perp (V,b_{|V})$ (see definition (2.1)). An element $x \in E$ is called strictly isotropic if $\langle x \rangle = Ax$ is a totally isotropic subspace of E. An element $z \in E$ is called strictly anisotropic if $\langle z \rangle = Az$ is a subspace of E with $q(z) \in A^*$, resp. $b(z,z) \in A^*$.

(3.2) Proposition. Let E be a quadratic resp. a bilinear module over A.

i) If E is non singular and $F \subseteq E$ is a subspace, then $F^{\perp}$ is a subspace and $F = F^{\perp\perp}$.

ii) Let $F \subseteq E$ be a submodule such that $(F,q_{|F})$ resp. $(F,b_{|F})$ is non singular. Then F is a subspace of E and $E = F \perp F^{\perp}$.

Proof. i) Let b be the bilinear form on E (in quadratic and bilinear case). Since F is a direct sumand of E, we have an exact sequence $E^* \to F^* \to o$, which together with $d_b : E \tilde{\to} E^*$ gives the exact sequence $d_b : E \to F^* \to o$. The kernel of $d_b : E \to F^*$ is exactly $F^{\perp}$, hence the exact sequence $o \to F^{\perp} \to E \to F^* \to o$. This sequence splits, since F is projectiv, i.e. $E \cong F^{\perp} \oplus F^*$, proving that $F^{\perp}$ is a subspace of E. Hence the isomorphism $d_b : E/F^{\perp} \tilde{\to} F^*$ induces a non singular pairing $b: F \times E/F^{\perp} \to A$, which applied to the subspace $F^{\perp}$ gives $b : F^{\perp} \times E/F^{\perp\perp} \to A$. On the other hand we have the isomorphism $E \cong (F^{\perp})^* \oplus F$ as a consequence of the isomorphisms $E^* \cong (F^{\perp})^* \oplus F^{**}$, $d_b : E \to E^*$ and $F^{**} \cong F$. From these facts we obtain the non singular pairing $b : F^{\perp} \times E/F \to A$. Using these last two pairings and the inclusion $F \subseteq F^{\perp\perp}$ we obtain $F = F^{\perp\perp}$.

ii) Let now $F \subseteq E$ be a submodule with $d_b : F \overset{\sim}{\to} F^*$, where E is any quadratic or bilinear module. Every element $x \in E$ defines an element in F^* by $y \to b(x,y)$ for all $y \in F$. Since $d_b : F \overset{\sim}{\to} F^*$, we obtain $z \in F$ with $b(z,y) = b(x,y)$ for all $y \in F$, i.e. $x - z \in F^{\perp}$. From $x = z + (x-z)$ it follows $E = F + F^{\perp}$. But $F \cap F^{\perp} = \{o\}$ since $d_b : F \to F^*$ is one to one, thus we have $E = F \perp F^{\perp}$. In particular F is a subspace of E.

(3.3) Corollary. Let E be a quadratic or bilinear module over A. Let a be an ideal contained in the Jacobson radical r of A. For every orthogonal decomposition $\bar{E} = \bar{F} \perp \bar{G}$ of $\bar{E} = E/aE$ over A/a, where $\bar{F}$ is a free non singular subspace of $\bar{E}$, there exists an orthogonal decomposition $E = F \perp G$ of E with F free and non singular, and $F/aF = \bar{F}$, $G/aG = \bar{G}$.

Proof. We set $\bar{F} = \langle\bar{x}_1\rangle \oplus \cdots \oplus \langle\bar{x}_n\rangle$ with $\bar{x}_i \in \bar{F}$ and $\det(\bar{b}(\bar{x}_i,\bar{x}_j)) \in (A/a)^*$ (see (1.2)). Choosing representatives $x_i \in E$ of the elements $\bar{x}_i \in \bar{F} \subseteq \bar{E}$ for $1 \leq i \leq n$ we define $F = Ax_1 + \cdots + Ax_n$. Then F is free and non singular. Because if $\lambda_1 x_1 + \cdots + \lambda_n x_n = o$ $(\lambda_i \in A)$, we obtain the n equations $b(x_1,x_i)\lambda_1 + \cdots + b(x_n,x_i)\lambda_n = o$, $1 \leq i \leq n$, and since $\det(b(x_i,x_j)) \in A^*$, it follows $\lambda_i = o$ for $i = 1,\ldots,n$. We now apply (3.2)(ii) and obtain $E = F \perp G$ with $G = F^{\perp}$. It is clear that $F/aF = \bar{F}$ and $G/aG = \bar{G}$.

Let us now consider the case of a semi-local ring A, i.e. a ring with only finitely many maximal ideals. The reduction of spaces over A with respect to the Jacobson radical of A leads us to consider spaces over fields. Now we shall recall some wellknown elementary results about quadratic forms over fields (see [A], [L], [W]). Let (E,q) be a quadratic space over the field k. If $ch(k) \neq 2$ we can find an orthogonal basis $\{x_1,\ldots,x_n\}$ for E with $q(x_i) \neq o$ and $(x_i,x_j) = o$ if $i \neq j$ (we shall often write (x,y) instead of $b_q(x,y)$ if the bilinear form b_q is fixed). In this case we have $E = \langle x_1\rangle \perp \ldots \perp \langle x_n\rangle = [c_1] \perp \ldots \perp [c_n]$ where $c_i = q(x_i)$. If $ch(k) = 2$, then $\dim E$ is even and there is a basis $\{e_1,f_1,\ldots,e_n,f_n\}$ of E with $\langle e_i,f_i\rangle \subseteq \langle e_j,f_j\rangle^{\perp}$ for $i \neq j$ and

$(e_i,f_i) = 1$ for all i. Let $q(e_i) = a_i$, $q(f_i) = b_i$, so that $<e_i,f_i> = [a_i,b_i]$. Then we have $E = [a_1,b_1]\perp\ldots\perp[a_n,b_n]$. In the case $ch(k) \neq 2$, $\dim E = 2n$, we can also find a basis $\{e_1,f_1,\ldots,e_n,f_n\}$ of E with $<e_i,f_i> \subseteq <e_j,f_j>^\perp$ for $i \neq j$ and $1 - 4q(e_i)q(f_i) \neq 0$, $(e_i,f_i) = 1$ for all i. With $q(e_i) = a_i$, $q(f_i) = b_i$ we obtain, as above, the decomposition $E = [a_1,b_1]\perp\ldots\perp[a_n,b_n]$. Let us now apply (3.3) to this remarks. We obtain

(3.4) Proposition. Let A be a semilocal ring and (E,q) be a free quadratic space over A.
Then E has an orthogonal decomposition

$$E = [a_1,b_1] \perp \ldots \perp [a_n,b_n]$$

or

$$E = [a_1,b_1] \perp \ldots \perp [a_n,b_n] \perp [c]$$

with $a_i,b_i \in A$, c, $1 - 4a_ib_i \in A^*$ $(1 \leq i \leq n)$ according as $\dim E = 2n$ or $2n+1$. If $2 \in A^*$ then E has an orthogonal basis, i.e.

$$E = [c_1] \perp \ldots \perp [c_m]$$

with $c_i \in A^*$, $1 \leq i \leq m$.

Proof. According to the preceding remarks, the proposition holds for fields and hence for finite direct products of fields. Now $A/r = \prod_m A/m$ is such a finite product of fields, so that our proposition follows directly from (3.3).

Considering bilinear spaces over semilocal rings we have a similar situation (see [M], [K]). We call a bilinear space (E,b) proper, if the ideal V(b) of A generated by all the values $b(x,x)$, $x \in E$, is the full ring A. Now it is easy to see that every proper bilinear space over a field has an orthogonal basis (see [M]). On the other hand a non proper bilinear space over a field is an orthogonal sum of copies of the space $\begin{pmatrix} 0 & 1 \\ 1 & 0 \end{pmatrix}$ (see [M]). Hence from (3.3) we deduce

(3.5) Proposition. Let (E,b) be a free bilinear space over the semi-local ring A.

i) If (E,b) is proper, then (E,b) has an orthogonal basis.

ii) If (E,b) is non proper, then (E,b) is an orthogonal sum of bilinear spaces $\begin{pmatrix} a & 1 \\ 1 & b \end{pmatrix}$ with $a,b \in r$.

We shall use in the sequel the following notation. For any $a \in A$ we denote the bilinear module (Ae,b) with $b(e,e) = a$ by $<a>$. An orthogonal sum $<a_1> \perp \ldots \perp <a_n>$ will be denoted by $<a_1,\ldots,a_n>$. The module $<a_1,\ldots,a_n>$ is non singular if and only if $a_i \in A^*$ for $1 \leq i \leq n$. Hence from (3.5) it follows that every proper bilinear space over a semi-local ring A has the form $<a_1,\ldots,a_n>$ with $a_i \in A^*$.

After this remarks about spaces over semilocal rings we return to the general case. Let A be a commutative ring and (E,q) be a quadratic space over A. The following result generalizes the Witt-Arf-decomposition of quadratic spaces over fields (see [A], [W], [Ba]).

(3.6) Theorem. Let $U \subset E$ be a totally isotropic subspace of E. Then there exist a totally isotropic subspace $W \subset E$ with $U \cong W^*$, so that $U \oplus W$ is a non singular subspace of E. Thus we have

$$E = U \oplus W \perp (U \oplus W)^{\perp}$$

Proof. Because $U^{\perp}$ is a subspace of E (see (3.2)) we have a direct sum decomposition $E = U^{\perp} \oplus V$. In the proof of (3.2) it has been shown that the pairing $b_q : U \times E/U^{\perp} \to A$ is non singular, i.e. $b_q : U \times V \to A$ is non singular. Hence we obtain the isomorphism $d_q : U \xrightarrow{\sim} V^*$ and $d_q : V \xrightarrow{\sim} U^*$. In particular it follows $U \cap V = \{o\}$, i.e. $U \oplus V$ is a non singular subspace of E (see (3.2)). If V is totally isotropic we are ready. If not, we shall change V to a totally isotropic subspace $W \subset E$ with the required properties. To this end we choose a bilinear form $b_0 : E \times E \to A$ with $q(x) = b_0(x,x)$ for all $x \in E$ (see (1.7)). Every $y \in V$ defines an element in V^* by $z \to b_0(y,z)$ for all $z \in V$. From the isomorphism $d_q : U \xrightarrow{\sim} V^*$ one obtains a unique element $u_y \in U$

with $d_q(y)(z) = b_o(y,z)$ for all $z \in V$, i.e. $b_q(u_y,z) = b_o(y,z)$. The uniqueness of u_y implies $u_{y+y'} = u_y + u_{y'}$ and $u_{\lambda y} = \lambda u_y$ for all $y,y' \in V$, $\lambda \in A$. Thus $W = \{y-u_y \mid y \in V\}$ is a submodule of E and we get a linear isomorphism $V \tilde{\to} W$ by $y \to y - u_y$. It can be easily be shown that $U \oplus W$ is a non singular subspace of E (see § 4). On the other hand W is totally isotropic, since for every $y \in V$ it holds $q(y-u_y) = q(y) - b_q(y,u_y) = q(y) - b_o(y,y) = o$. This proves our assertion.

(3.7) Remark. For bilinear spaces over A there is a similar result: let (E,b) be a bilinear space and $U \subset E$ a totally isotropic subspace. Then there exists a subspace $V \subset E$ with $U \tilde{\to} V^*$, so that $U \oplus V$ is a non singular subspace of E. The proof of this fact is the same as in the quadratic case, though it is not true in general, that V can be choosen totally isotropic. For example let k be a field with ch(k) = 2 and $E = ke \oplus kf$, where $b(e,e) = o$, $b(e,f) = 1$, $b(f,f) = 1$. One sees easily that $\langle e \rangle$ is the only totally isotropic subspace of E.

(3.8) Corollary. Let E be a quadratic or bilinear space over A. Let $U \subset E$ be a totally isotropic subspace of maximal rank, i.e. $U = U^{\perp}$. Then there is a subspace $W \subset E$ with $U \cong W^*$ and $E = U \oplus W$. If E is a quadratic space we can choose W totally isotropic. In particular $r_E = 2r_u$.

Proof. According to (3.6) and (3.7) we can find a subspace $W \subset E$ with $U \cong W^*$ and $U \oplus W$ non singular subspace of E. Moreover if E is quadratic space W can be choosen totally isotropic. We have $r_U = r_W$. On the other hand we have a decomposition $E = U^{\perp} \oplus V$ with some $V \cong W$ (see proof of (3.7)). Since $U = U^{\perp}$, it follows $r_E = 2r_U$ and $E = U \oplus W$. This proves the corollary.

(3.9) Example. If (E,q) is a two dimensional quadratic space and $x \in E$ is strictly isotropic, then we can find an strictly isotropic element $y \in E$ with $E = \langle x \rangle \oplus \langle y \rangle$, and without restriction $(x,y) = 1$, i.e. $E = [o,o]$. More generally let E be a free 2n-dimensional quadratic space which contains a free n-dimensional totally isotropic subspace U. Therefore $U = U^{\perp}$. Let $U = Ax_1 \oplus \cdots \oplus Ax_n$. Using the isomorphism $d_q : E \to E^*$ we get n elements $x_1^*,\dots,x_n^* \in E$ with $(x_i,x_j^*) = \delta_{ij}$ for all i,j. In this way we obtain a subspace $V = Ax_1^* \oplus \cdots \oplus Ax_n^*$ with the property $E = U \oplus V$. As in the proof of (3.6) we can replace V by a totally isotropic subspace W using a suitable isomorphism $V \tilde{\to} W$. Then $E = U \oplus W$. Let $y_1,\dots,y_n$ be the images of the $x_1^*,\dots,x_n^*$ in W under this isomorphism. Then it holds $(x_i,y_i) = \delta_{ij}$ for all i,j, and we get in this manner $E = \langle x_1,y_1\rangle \perp \dots \perp \langle x_n,y_n\rangle = [o,o] \perp \dots \perp [o,o]$. The next section will be concerned with the study of this kind of spaces.

§ 4. Hyperbolic spaces

Take $P \in \underline{P}(A)$. We define on $P \oplus P^*$ a quadratic form q_P by

$$(4.1) \qquad q_P(x + x^*) = x^*(x)$$

for all $x \in P$, $x^* \in P^*$. The associated bilinear form is given by $b_P(x + x^*, y + y^*) = x^*(y) + y^*(x)$ for all $x,y \in P$, $x^*,y^* \in P^*$. In (4.4) below it is shown that $(P \oplus P^*, q_P)$ is non singular. The quadratic space $(P \oplus P^*, q_P)$ will be called the _hyperbolic space_ associated to $P \in \underline{P}(A)$ and it will be denoted by $\mathbb{H}[P]$. In a similar way we can construct hyperbolic bilinear spaces. Indeed let (U,b) be a bilinear module over A. We define on the module $U \oplus U^*$ a symmetric bilinear form b_U by

$$(4.2) \qquad b_U(u + u^*, v + v^*) = b(u,v) + u^*(v) + v^*(u)$$

for all $u,v \in U$, $u^*,v^* \in U^*$. The space (see (4.4)) $(U \oplus U^*, b_U)$ is called the _metabolic space_ associated with (U,b). These spaces were introduced by Knebusch in $[K]_1$. The metabolic space $(U \oplus U^*, b_U)$ will be denoted by $\mathbb{M}(U)$, but in the special case b = o we shall write $\mathbb{H}(U)$ instead of $\mathbb{M}(U)$. The space $\mathbb{H}(U)$ is also called the _hyperbolic bilinear space_ associated to $U \in \underline{P}(A)$.

(4.3) <u>Remarks</u>. 1) In $\mathbb{H}[P]$ there are two totally isotropic subspaces, namely P and P^*, so that $P = P^{\perp}$, $P^* = P^{*\perp}$. In $\mathbb{M}(U)$ in general only U^* is a totally isotropic subspace, because U is endowed with a bilinear form.

2) If $P = Ae$ is onedimensional, then $\mathbb{H}[Ae] = [o,o]$. Similarly for the bilinear module $\langle a \rangle$ $(a \in A)$ we have $\mathbb{M}(\langle a \rangle) = \begin{pmatrix} a & 1 \\ 1 & o \end{pmatrix}$. More generally, if $P = Ae_1 \oplus \cdots \oplus Ae_n$, then

$$\mathbb{H}[P] = \begin{bmatrix} o & I_n \\ I_n & o \end{bmatrix} \quad \text{with} \quad I_n = \begin{pmatrix} 1 & & o \\ & \ddots & \\ o & & 1 \end{pmatrix}$$

i.e. $\mathbb{H}[P] = [o,o] \perp \ldots \perp [o,o]$ (n-times). In the sequel we often shall denote the hyperbolic plane $[o,o]$ by $\mathbb{H}$. Analogously we have for the bilinear module $(U,b) = \langle a_1,\ldots,a_n \rangle$ $(a_i \in A)$

$$\mathbb{M}(\langle a_1,\ldots,a_n \rangle) = \begin{pmatrix} a_1 & 1 \\ 1 & o \end{pmatrix} \perp \ldots \perp \begin{pmatrix} a_n & 1 \\ 1 & o \end{pmatrix}$$

(4.4) <u>Proposition</u>. i) For every $P \in \underline{P}(A)$ is $\mathbb{H}[P]$ a non singular quadratic module.

ii) For every bilinear module (U,b) is $\mathbb{M}(U)$ a non singular bilinear module.

<u>Proof</u>. We only treat the quadratic case, because the bilinear case is a similar one. From $\mathbb{H}[P] = P \oplus P^*$ follows that $\mathbb{H}[P]^* = P^* \oplus P^{**}$. The associated linear form $d_P : \mathbb{H}[P] \to \mathbb{H}[P]^*$ to q_P has, with respect to the above decompositions, the matrix

$$d_P = \begin{pmatrix} o & j_P \\ id_{P^*} & o \end{pmatrix}$$

where $j_P : P \xrightarrow{\sim} P^{**}$ is the canonical isomorphism. This follows from $d_P(u) = j_P(u)$ and $d_P(u^*) = u^*$ for all $u \in P$ and $u^* \in P^*$. Hence d_P is an isomorphism.

The proof of the next proposition is straightforward, so that we omit it.

(4.5) <u>Proposition</u>. For $P,Q \in \underline{P}(A)$ it holds

$$\mathbb{H}[P \oplus Q] \cong \mathbb{H}[P] \perp \mathbb{H}[Q]$$

and for any bilinear module (U,b) with a decomposition $U = U_1 \perp U_2$

$$\mathbb{M}(U) \cong \mathbb{M}(U_1) \perp \mathbb{M}(U_2)$$

Our next aim is to characterize the hyperbolic and metabolic spaces respectively. This we perform in the next proposition.

(4.6) <u>Theorem</u>. Let E be a quadratic or bilinear space over A. Then E is a hyperbolic, resp. metabolic space, if and only if E contains a totally isotropic subspace U with $U = U^{\perp}$. Moreover in the quadratic case $E \cong \mathbb{H}[U]$.

<u>Proof</u>. Let (E,q) be a quadratic space with a subspace $U \subset E$, so that $q(U) = o$, $U = U^{\perp}$. According (3.8) one can find a subspace $W \subset E$ with $q(W) = o$, $d_q : W \tilde{\to} U^*$ and $E = U \oplus W$. Defining $f : E \to \mathbb{H}[U] = U \oplus U^*$ by $f(u+w) = u + d_q(w)$ for $u \in U$, $w \in W$ we obviously obtain a linear isomorphism of modules, which is also an isomorphism of quadratic spaces, since

$$q_U(f(u+w)) = q_U(u + d_q(w)) = d_q(w)(u)$$

$$= b_q(u,w) = q(u+w)$$

This proves our assertion in the quadratic case. Now the bilinear case can be treated similarly.

(4.7) <u>Corollary</u>. (i) For $(E,q) \in \underline{\text{Quad}}(A)$ it holds

$$(E,q) \perp (E,-q) \cong \mathbb{H}[E]$$

ii) For $(U,b) \in \underline{\text{Bil}}(A)$ it holds

$$(U,b) \perp (U,-b) \cong \mathbb{M}(U)$$

iii) For $P \in \underline{P}(A)$, $(E,q) \in \underline{Quad}(A)$, $(M,b) \in \underline{Bil}(A)$ and (U,b') any bilinear module we have

1) $\mathbb{M}(U) \otimes (E,q) \cong \mathbb{H}[U \otimes E]$

2) $(M,b) \otimes \mathbb{H}[P] \cong \mathbb{H}[M \otimes P]$

3) $\mathbb{H}[P] \circ (E,q) \cong \mathbb{H}[P \otimes E]$

4) $\mathbb{M}(U) \otimes (M,b) \cong \mathbb{M}(U \otimes M)$

Proof. (i) As $E \perp -E$ contains the totally isotropic subspace $V = \{(x,x) \mid x \in E\}$ ($\cong E$ as modules) with $V = V^{\perp}$, we obtain (i) immediately from (4.6). The case (ii) can be proved similarly.

iii) The quadratic space $\mathbb{M}(U) \otimes E$ contains the maximal totally isotropic subspace $U^* \otimes E$, thus we obtain $\mathbb{M}(U) \otimes E \cong \mathbb{H}[U^* \otimes E] \cong \mathbb{H}[U \otimes E]$. The remaining formulas can be proved in a similar way. We omit the details.

Now let $\hat{W}_q(A)$ and $\hat{W}(A)$ be the Witt-Grothendieck rings of quadratic and bilinear spaces over A respectively. We define

$$\hat{\mathbb{H}}(A) = \{[\mathbb{H}[P]] - [\mathbb{H}[Q]] \mid P,Q \in \underline{P}(A)\}$$

and

$$\hat{\mathbb{M}}(A) = \{[\mathbb{M}(U)] - [\mathbb{M}(V)] \mid U,V \text{ bilinear modules}\}$$

The corollary (4.7) asserts that $\hat{\mathbb{H}}(A)$ and $\hat{\mathbb{M}}(A)$ are ideals in $\hat{W}_q(A)$ and $\hat{W}(A)$ respectively. We also obtain the relations $\hat{W}(A)\ \hat{\mathbb{H}}(A) \subseteq \hat{\mathbb{H}}(A)$ and $\hat{\mathbb{M}}(A)\ \hat{W}_q(A) \subseteq \hat{\mathbb{H}}(A)$. Now we are in condition to make the following definition.

(4.8) Definition. We set

$$W_q(A) = \hat{W}_q(A) \ / \ \hat{\mathbb{H}}(A)$$

$$W(A) = \hat{W}(A) \ / \ \hat{\mathbb{M}}(A)$$

$W_q(A)$ is called the <u>Witt ring of quadratic spaces</u> over A and W(A) is called the <u>Witt ring of bilinear spaces</u> over A. $W_q(A)$ is a W(A)-algebra but in general without 1. The general element in $W_q(A)$ has the form [E] - [F] with quadratic spaces E and F. Now from (4.7) and the defining relations of $W_q(A)$ we obtain in $W_q(A)$

$$[E] - [F] = [E] + [-F] - ([F] + [-F])$$

$$= [E \perp -F] \ ,$$

that is the elements in $W_q(A)$ are represented by classes [E] of quadratic spaces. In $W_q(A)$ we have [E] = [F] if and only if there are modules $P, Q \in \underline{P}(A)$ with $E \perp \mathbb{H}[P] \cong F \perp \mathbb{H}[Q]$. In this case we shall write $E \sim F$. In the bilinear case the situation is similar, so we omit the details.

Every ring homomorphism $\alpha : A \to B$ induces ring homomorphisms $\alpha^* : \hat{W}_q(A) \to \hat{W}_q(B)$ and $\alpha^* : \hat{W}(A) \to \hat{W}(B)$ with the properties $\alpha^*[\hat{\mathbb{H}}(A)] \subseteq \hat{\mathbb{H}}(B)$ and $\alpha^*[\hat{\mathbb{M}}(A)] \subseteq \hat{\mathbb{M}}(B)$ respectively. This leads us to define the homomorphisms of rings

$$\alpha^* : W_q(A) \to W_q(B) \text{ and } \alpha^* : W(A) \to W(B)$$

Let now $A \xrightarrow{i} B$ be a Frobenius extension with the trace map $s : B \to A$. As above we obtain two homomorphisms of groups

$$s_* : W_q(B) \to W_q(A) \quad \text{and} \quad s_* : W(B) \to W(A)$$

which we shall call the <u>transfer homomorphisms</u> defined by $s : B \to A$. Naturally the formulas in (2.12) are still true forthese homomorphisms.

(4.9) <u>Example</u>. Let A be a Dedekind ring with quotient field K_A. The inclusion $i : A \to K_A$ induces the homomorphisms $i^* : W_q(A) \to W_q(K_A)$ and $i^* : W(A) \to W(K_A)$. Now we want to show that they are injectiv. In fact we have the following result (see $[K]_1$).

(4.10) Proposition. Let E be a quadratic or bilinear space over A, so that $E \otimes K_A$ is hyperbolic or metabolic, respectively. Then E is hyperbolic or metabolic over A, respectively.

Proof. Let $E \otimes K_A = U' \oplus V'$ with $(U')^{\perp} = U'$, and $q(U') = o$ in the quadratic case. We set $U = E \cap U'$ and obtain an exact sequence $o \to U \to E \to E/U \to o$. Observe that E/U is a finitely generated torsionfree module over A, hence E/U is projectiv. It follows that the sequence splits, i.e. $E \cong U \oplus E/U$, which implies that U is a subspace of E. Clarly $U = U^{\perp}$ and $q(U) = o$. The assertion follows from (4.6).

(4.11) Corollary. For every Dedekind ring A the inclusion $i : A \to K_A$ induces monomorphisms $i^* : W_q(A) \to W_q(K_A)$ and $i^* : W(A) \to W(K_A)$.

Now over every ring A there is a canonical homomorphism of rings $\beta : W_q(A) \to W(A)$ defined by $\beta([q]) = [b_q]$ for all $[q] \in W_q(A)$. If A is a Dedekind ring we obtain a commutative diagram

$$\begin{array}{ccc} W_q(K_A) & \xrightarrow{\beta} & W(K_A) \\ i^* \uparrow & & \uparrow i^* \\ W_q(A) & \xrightarrow{\beta} & W(A) \end{array}$$

where the vertical rows are monomorphisms. Since $\beta : W_q(K_A) \to W(K_A)$ is an isomorphisms if $2 \neq o$ in A, we deduce

(4.12) Corollary. For every Dedekind ring A with $2 \neq o$

$$\beta : W_q(A) \to W(A)$$

is a monomorphism.

(4.13) Example. As an application of this results we shall compute $W_q(\mathbb{Z})$ (compare [M-H] for the computation of $W(\mathbb{Z})$). The inclusion $j : \mathbb{Z} \to \mathbb{R}$ induces a homomorphism $j^* : W(\mathbb{Z}) \to W(\mathbb{R})$, and since $W(\mathbb{R}) \cong \mathbb{Z}$ (Sylvester law of inertia), we obtain a homomorphism

$j^*: W(\mathbb{Z}) \to \mathbb{Z}$ with $j^*(<1>) = 1$. In order to prove that j^* is an isomorphism, we only need to check the injectivity of j^*. Let $[b] \in W(\mathbb{Z})$ be an element with $j^*[b] = [b \otimes \mathbb{R}] = o$. This means that $b \otimes \mathbb{R}$ is totally indefinite and dim b is even. Let b be defined on the module E. If dim E $\geq$ 6, then $b \otimes Q$ is isotropic in virtue of the theorem of Meyer (see [Se], S.77). Therefore there is an $x \in E$, $x \neq o$ with $b(x,x) = o$. Without restriction we can assume x to be undivisible, which implies that $<x> = \mathbb{Z}x$ is a totally isotropic subspace of (E,b). Hence $E = <x,y> \perp E_o$ with $<x,y>$ metabolic (see (3.7)). In consequence we can suppose dim E $\leq$ 4. But det b = ±1 because (E,b) is non singular over $\mathbb{Z}$ and since $b \otimes \mathbb{R}$ is totally indefinite, we obtain $\det(b \otimes Q) = -1$ if dim E = 2 and $\det(b \otimes Q) = 1$ if dim E = 4. In the case dim E = 2 follows immediately $[b \otimes Q] = o$ in W(Q), and hence $[b] = o$ in $W(\mathbb{Z})$ (see (4.10)). Assume now dim E = 4. Then $b \otimes Q$ has the form $\sum a_{ij}X_iX_j$ with $a_{ij} \in \mathbb{Z}$ and $\det(a_{ij}) = 1$. Taking reduction modulo a prime $p \neq 2$ we get an isotropic form over $\mathbb{Z}/(p)$, and from this follows that $b \otimes Q_p$ is isotropic for all primes $p \neq 2$. Since $b \otimes Q$ is totally indefinite and $\det(b \otimes Q) = 1$, we can apply Cor.3, chap.IV, §3 in [Se] to deduce, that $b \otimes Q$ is isotropic over Q. Again we obtain $E = <x,y> \perp E_o$ over $\mathbb{Z}$ with $<x,y>$ metabolic, $\dim E_o = 2$. This implies $[b] = o$. Hence we have proved $j^*: W(\mathbb{Z}) \xrightarrow{\sim} \mathbb{Z}$. Using the monomorphism $\beta: W_q(\mathbb{Z}) \to W(\mathbb{Z})$ we obtain a monomorphism $j^* \circ \beta: W_q(\mathbb{Z}) \to \mathbb{Z}$. It is easy to show that every quadratic space (E,q) over $\mathbb{Z}$ with dim E $\leq$ 6 is hyperbolic (see (4.24), chap.V). On the other hand we have over $\mathbb{Z}$ the anisotropic quadratic space E_8, whose value matrix is (see [Se], S.89).

$$E_8 = \begin{bmatrix} 1 & o & -1 & o & o & o & o & o \\ o & 1 & o & -1 & o & o & o & o \\ -1 & o & 1 & -1 & o & o & o & o \\ o & -1 & -1 & 1 & -1 & o & o & o \\ o & o & o & -1 & 1 & -1 & o & o \\ o & o & o & o & -1 & 1 & -1 & o \\ o & o & o & o & o & -1 & 1 & -1 \\ o & o & o & o & o & o & -1 & 1 \end{bmatrix}$$

Hence $[E_8] \neq o$ in $W_q(\mathbb{Z})$ and it follows $j^* \circ \beta([E_8]) = 8$ in $\mathbb{Z}$. Thus we obtain $j^* \circ \beta: W_q(\mathbb{Z}) \xrightarrow{\sim} \mathbb{Z}\cdot 8$.

CHAPTER II

Invariants of quadratic forms

§ 1. Azumaya Algebras.

In this section we shall collect some known results over separable algebras, whose proofs can be found in the current literature (see for example [A-G], [Ba], [DeM-I]). In the sequel A will denote always a commutative ring with 1 and all considered modules over A shall belong to $\underline{P}(A)$. We remind the reader of the definition of Azumaya-algebras (see [Ba], S.104): an A-algebra B is called Azumaya-algebra if it has the following properties

i) $B \in \underline{P}(A)$ as A-module

ii) for every $m \in \max(A)$ is B/mB a central simple algebra over the field A/m.

Equivalently we have: the A-algebra B is an Azumaya-algebra if and only if

(i)' B is a finitely generated A-module

(ii)" B is a central separable A-algebra,

where separable algebras over rings are defined as follows: let B be an A-algebra and B^o be the opposite algebra. B is called separable over A if and only if B is a projective $B^e = B \otimes_A B^o$-module. An important local characterization of separable A-algebras is the following result (see [Ba], [deM-I]).

(1.1) Proposition. Let B be an A-algebra which is finitely generated as A-module. Then the following assertions are equivalent:

i) B is separable over A

ii) B_m is separable over A_m for all $m \in \max(A)$

iii) B/mB is separable over A/m for all $m \in \max(A)$.

For example if $P \in \underline{P}(A)$, then $\mathrm{End}_A(P)$ is a separable A-algebra, whose center is A/ann(P). If P is a faithfully projective A-module, then End(P) is an Azumaya-algebra. For Azumaya algebras we have the following characterization.

(1.2) Proposition. Let B be an A-algebra. If B is an Azumaya-algebra, then B is a faithfully projective A-module and $B \otimes_A B^o \cong End_A(B)$. Conversely if there exists an A-algebra C and a faithfully projective A-module P with $B \otimes_A C \cong End_A(P)$, then B is an Azumaya-algebra.

We denote the set of isomorphism classes [P] of projective A-modules of rank one by Pic(A). On Pic(A) is a product defined by $[P_1]\cdot[P_2] = [P_1 \otimes_A P_2]$, and Pic(A) with this product is an abelian group of exponent two (see $[Bo]_1$). For example if A is semilocal, then Pic(A) = o. With every A-automorphism σ of an Azumaya-algebra B we can associate an element of Pic(A) in the following way: it is easy to see that the A-module $I_\sigma = \{b \in B \mid \sigma(x)b = bx' \text{ for all } x \in B\}$ is projective of rank one, hence one obtain a map $Aut_A(B) \to Pic(A)$ by $\sigma \to [I_\sigma]$. The kernel of this homomorphism is the subgroup $Inn_A(B)$ of inner automorphism of B, i.e. we have an exact sequence

$$(1.3) \qquad 1 \to Inn_A(B) \to Aut_A(B) \to Pic(A)$$

which generalizes the classical theorem of Skolem-Noether.

Let $\underline{Az}(A)$ be the category of Azumaya-algebras over A. It is easy to prove that if $B_1, B_2 \in \underline{Az}(A)$, then $B_1 \otimes_A B_2 \in \underline{Az}(A)$. On $\underline{Az}(A)$ we define an equivalence relation $\sim$ by specifying that two elements $B_1, B_2 \in \underline{Az}(A)$ are similar (written $B_1 \sim B_2$) if and only if there exist faithfully projective A-modules $P_1, P_2 \in \underline{P}(A)$ with

$$B_1 \otimes_A End_A(P_1) \cong B_2 \otimes_A End_A(P_2)$$

Equivalently we have $B_1 \sim B_2$ if and only if

$$B_1 \otimes B_2^o \cong End_A(P)$$

for some faithfully projective $P \in \underline{P}(A)$. Now we define the Brauer group of the ring A. Let Br(A) be the set of equivalence classes [B] of Azumaya-algebras over A with respect to the relation $\sim$. We denote the equivalence class [End(P)] for all faithfully projective A-modules P by 1 and the class $[B^o]$ by $[B]^{-1}$. Let us introduce in Br(A) a product in the following way: for any $[B_1]$, $[B_2] \in Br(A)$ set

$$[B_1]\cdot[B_2] = [B_1 \otimes_A B_2]$$

Then it can easily be seen that Br(A) is an abelian group with identity 1. The inverse of [B] is $[B]^{-1} = [B^o]$. This group is called the Brauer group of A.

We now introduce another class of algebras over A, which shall play an important role by defining invariants of quadratic forms.

(1.4) Definition. A separable A-algebra B is called a quadratic separable A-algebra if B is a finitely generated A-module of rank 2.

Let B be a quadratic separable A-algebra. From (1.1) follows that B/mB is either a quadratic separable extension of A/m or the algebra $A/m \times A/m$ for every $m \in \max(A)$. This fact implies immediately that A is a direct sumand of B, and B is commutative (see (2.9), (2.17), chap. III in [Ba]). Every quadratic separable A-algebra B posses a unique automorphism σ_B with $\sigma_B{}^2 = id_B$ and $Fix(\sigma_B) = \{b \in B \mid \sigma_B(b) = b\} = A$ (see [Ba]). Since we shall be later mainly concerned with free modules, we shall determine explicitely the form of a free quadratic separable algebra over A (see [Ra]). A monic polynomial $f(t) \in A[t]$ is called separable if $\overline{f}(t) \in A/m[t]$ is separable for all $m \in \max(A)$. For example $f(t) = t^2 + at + b$ is separable if and only if $a^2 - 4b \in A^*$. This polynomials characterize all free quadratic separable algebras over A. In fact we have

(1.5) Theorem. Let B be a free quadratic separable algebra over A. Then there exist a separable polynomial $t^2 - at - b \in A[t]$ such that

$$B \cong A[t]\,/\,(t^2 - at - b)$$

Proof. Since A is a direct sumand of B, we have $B = A \oplus C$ for a submodule C of B. On the other hand $B \cong A \oplus A$, which implies $A \cong \Lambda^2 B \cong C$, that is $C = A.z$ is free ($z \in B$). Thus $B = A \oplus Az$, and there are $a, b \in A$ with $z^2 = az + b$. Therefore we obtain a homomorphism of algebras

$$A[t]/(t^2 - at - b) \to B$$

which is defined by $t \to z$. This homomorphism is onto, since $B \cong A \oplus Az$. Compairing the ranks of both algebras we deduce that it is one to one, that is $B \cong A[t]/(t^2 - at - b)$. In particular $t^2 - at - b$ must be separable.

If A is a semilocal ring we can give a more precise description of such algebras, namely

(1.6) Theorem. Let A be a semilocal ring. Then any quadratic separable A-algebra B has the form

$$B = A[t]/(t^2 - t - b)$$

with $1 + 4b \in A^*$.

Proof. In virtue of (1.5) we can write $B = A \oplus Az$ with $z^2 = az + b$ and $a^2 + 4b \in A^*$. We shall change the basis $\{1,z\}$ of B to a new basis $\{1,w\}$, so that $w^2 = w + c$ and $1 + 4c \in A^*$. To this end let us set $w' = z + d$ with some $d \in A$. Then $w'^2 = (a + 2d)w' + b - ad - d^2$. We choose d, so that $a + 2d \in A^*$. This is always possible because we can use the chinese remainder theorem to solve the congruences $d \equiv 1 \pmod{m}$ if $a \in m$ and $d \equiv o \pmod{m}$ if $a \notin m$ for all $m \in \max(A)$. Now we define $w = (a + 2d)^{-1}w'$. From the above relation for w' follows $B = A \oplus Aw$, $w^2 = w + c$ with a suitable c such that $1 + 4c \in A^*$.
Over any commutative ring A the unique involution σ_B of $B = A \oplus Az$, $z^2 = az + b$ $(a^2 + 4b \in A^*)$ is given by $\sigma_B(z) = a - z$. We shall denote the quadratic separable algebra $A \oplus Az$ with $z^2 = z + b$ $(1 + 4b \in A^*)$ by $A(\wp^{-1}(b))$. Thus over a semilocal ring every quadratic separable algebra has this form. Let us denote the set of isomorphism classes of quadratic separable algebras over A by $\Delta(A)$. This set can be endwed with the structure of an abelian group of exponent 2 in the following way: let B_1, B_2 be two quadratic separable algebras over A with the canonical involutions σ_1 and σ_2 respectively. We have on $B_1 \otimes B_2$ the involution $\sigma_1 \otimes \sigma_2$. We define

(1.7) $$B_1 \circ B_2 = \mathrm{Fix}(\sigma_1 \otimes \sigma_2)$$

It can easily be seen that $B_1 \circ B_2$ is a quadratic separable algebra over A, whose canonical involution is given by the restriction of $\sigma_1 \otimes \mathrm{id}_{B_2}$ (or $\mathrm{id}_{B_1} \otimes \sigma_2$) to $B_1 \circ B_2$. Let us take for example $B_1 = A \oplus Az_1$, $B_2 = A \oplus Az_2$ with $z_i^2 = z_i + b_i$, $1 + 4b_i \in A^*$ $(i = 1,2)$. An easy calculation shows that $B_1 \circ B_2 = A \oplus Az$ with $z = z_1 \otimes 1 + 1 \otimes z_2 - 2z_1 \otimes z_2$ and $z^2 = z + b_1 + b_2 + 4b_1b_2$, that is

(1.8) $$A(\wp^{-1}(b_1)) \circ A(\wp^{-1}(b_2)) = A(\wp^{-1}(b_1+b_2+4b_1b_2)).$$

With this product $\Delta(A)$ becomes an abelian group of exponent 2. The unit element is defined by the class $1 = [A \times A]$, and since $B \circ B \cong A \times A$ for every quadratic separable algebra B, we obtain $[B]^2 = 1$. The group $\Delta(A)$ will be called the group of quadratic separable extensions of A. Both $\Delta(A)$ and $\mathrm{Br}(A)$ will play an important role in the study of invariants of quadratic forms.

(1.11) Remark. Let A be a semilocal ring. If $2 \in A^*$ we get an isomorphism

$$\Delta(A) \xrightarrow{\sim} A^*/A^{*2}$$

which is defined by $[A(\wp^{-1}(b))] \to (1 + 4b) \bmod A^{*2}$. If $4 = 0$ in A, then the algebra $A(\wp^{-1}(b))$ is separable for every $b \in A$. According to (1.10) we obtain an homomorphism

$$\alpha : A \to \Delta(A)$$
$$b \to [A(\wp^{-1}(b))]$$

which obviously is onto. It follows at once from the definitions that $\mathrm{Ker}(\alpha) = \{a^2 - a \mid a \in A\}$, since $A(\wp^{-1}(b)) \cong A \times A$ if and only if $b = a^2 - a$ for some $a \in A$. We shall denote the subgroup $\{a^2 - a \mid a \in A\}$ by $\wp(A)$. Using the relation $a^2-a+b^2-b = (1-a-b+2ab)^2 - (1-a-b+2ab)$ for all $a,b \in A$ it can be seen directly that $\wp(A)$ is a subgroup of A.

Hence $\Delta(A) \cong A/\wp(A)$.
Let now B be a quadratic separable algebra over any commutative ring A and let $a \in A^*$ be any unit of A. We construct the module $D = B \oplus Be$ (free over B) endowed with a structure of A-algebra defined by $e^2 = a$, $xe = e\sigma_B(x)$ for all $x \in B$, where σ_B is the canonical involution of B. For example if $B = A(\wp^{-1}(b)) = A \oplus Az$ with $z^2 = z + b$ $(1 + 4b \in A^*)$, then D is free over A with basis $\{1, z, e, ze\}$ and the relations $z^2 = z + b$, $e^2 = a$, $ze + ez = e$. Hence in the general case D is an Azumaya-algebra over A which we shall denote by $(a,B]$. The algebra $(a,B]$ is called the <u>quaternion algebra</u> associated to $a \in A^*$ and B. In case $B = A(\wp^{-1}(b))$ we shall write $(a,b]$ instead of $(a,B]$. It is easy to see that B is a maximal commutative subalgebra of $(a,B]$ and $(a,B] \otimes_A B \cong \mathrm{End}_B((a,B])$, i.e. B is a splitting ring for $(a,B]$. We now list below some elementary relations which are fulfilled by the elements $[(a,B]] \in Br(A)$ for all $a \in A^*$, $[B] \in \Delta(A)$.

(1.12) $$[(a^2,B]] = 1, \quad [(a, A \times A]] = 1$$

(1.13) $$[(ac,B]] = [(a,B]] \cdot [(c,B]]$$

(1.14) $$[(a, B_1 \circ B_2]] = [(a,B_1]] \cdot [(a,B_2]]$$

In particular it follows that the symbol $[(a,B]]$ depends only on the square class $a \pmod{A^{*2}}$ in A^*/A^{*2}, and the isomorphism class $[B] \in \Delta(A)$. In this way we obtain a bimultiplicative pairing

(1.15) $$A^*/A^{*2} \times \Delta(A) \to Br(A)_2$$

The image of this pairing lies in $Br(A)_2$ because

(1.16) $$[(a,B]]^2 = 1.$$

All this facts can easily be obtained from the following isomorphism

(1.17) $$(a_1 a_2, B_2] \otimes (a_1, B_1 \circ B_2] \xrightarrow{\sim} (a_1, B_1] \otimes (a_2, B_2]$$

for $a_1, a_2 \in A^*$, $[B_1], [B_2] \in \Delta(A)$.

In order to establish the fundamental relationships between quadratic forms and these algebras, we introduce the norm form of quadratic separable algebras and quaternion algebras.

First let B be a quadratic separable algebra over A with the canonical involution σ_B. We define the norm map $n : B \to A$ by $n(x) = x\sigma_B(x)$ for all $x \in B$. Then n defines on $B \in \underline{P}(A)$ a non singular quadratic form over A, whose associated bilinearform is $b_n(x,y) = x\,\sigma_B(y) + y\,\sigma_B(x)$.

The quadratic space (B,n) is called the norm form of B.

If $B = A(\wp^{-1}(b)) = A \oplus Az$ with $z^2 = z + b$, then the matrix of (B,n) with respect to {1,z} is

$$\begin{bmatrix} 1 & 1 \\ 1 & -b \end{bmatrix}$$

i.e. (B,n) = [1,-b]. This quadratic space will often be denoted by [B]. It should be noted that this notation does not contradicts the notation $[B] \in \Delta(A)$ because of the following result.

(1.18) Proposition. Let B_1, B_2 be two quadratic separable algebras over A. Then $B_1 \cong B_2$ if and only if $(B_1,n_1) \cong (B_2,n_2)$.

Proof. First we prove that if $B_1 \cong B_2$ then $(B_1,n_1) \cong (B_2,n_2)$. The converse implication will be proved in § 3. Let $\varphi : B_1 \to B_2$ be an isomorphism. From the uniqueness of the canonical involutions we deduce $\varphi(\sigma_1(x)) = \sigma_2(\varphi(x))$ for all $x \in B_1$. It follows for every $x \in B_1$

$$\begin{aligned} n_1(x) &= \varphi(n_1(x)) = \varphi(x\sigma_1(x)) = \varphi(x)\varphi(\sigma_1(x)) \\ &= \varphi(x)\sigma_2(\varphi(x)) = n_2(\varphi(x)) \end{aligned}$$

that is $\varphi : B_1 \xrightarrow{\sim} B_2$ defines an isomorphism $\varphi : (B_1,n_1) \xrightarrow{\sim} (B_2,n_2)$, as was to be shown.

Let us now consider a quaternion algebra D = (a,B] over A with $a \in A^*$, $[B] \in \Delta(A)$. The reduced norm map of $D = B \oplus Be$ $(e^2 = a)$ is given by

$$\bar{n} : D \to A$$

$$\bar{n}(u+ve) = n(u) - an(v)$$

for $u,v \in B$, where $n:B \to A$ is the norm map of B. It is easy to check that $\bar{n}$ is multiplicative, i.e. $\bar{n}(xy) = \bar{n}(x)\bar{n}(y)$ for all $x,y \in D$. Now n defines on $D \in \underline{P}(A)$ a non singular quadratic form (of rank 4), which we shall denote by [D]. The quadratic space $[D] = (D,\bar{n})$ is called the <u>norm form</u> of D. For example if $D = (a,B]$ then $[D] = \langle 1,-a\rangle \otimes [B]$, and in the special case $B = A(\wp^{-1}(b))$ we have $[D] = \langle 1,-a\rangle \otimes [1,-b]$. Our notation [D] for $(D,\bar{n})$ is not full consistent with the notation $[D] \in Br(A)$, as (1.19) below shows, but we maintain it, since it does not lead to any complication.

(1.19) <u>Proposition</u>. Let A be any ring and let D_1,D_2 be two quaternion algebras over A. If $D_1 \cong D_2$ then $(D_1,\bar{n}_1) \cong (D_2,\bar{n}_2)$. Moreover, if for the ring A the Azumaya algebras are determined (up to isomorphism) by their class in Br(A) and the rank, then $(D_1,\bar{n}_1) \cong (D_2,\bar{n}_2)$ implies $D_1 \cong D_2$ (this last condition is always satisfied for semilocal rings).

<u>Proof</u>. We prove the first assertion and postpone the proof of the second to § 3. Let $\varphi : D_1 \xrightarrow{\sim} D_2$ be an isomorphism of A-algebras. From the uniqueness of the reduced norm (see [Kn-O], ch.IV, §2) follows $\bar{n}_1(x) = \bar{n}_2(\varphi(x))$ for all $x \in D_1$, which proves that φ is an isomorphism of quadratic spaces.

(1.20) <u>Remark</u>. An immediately consequence of this results are the following facts.

i) Let $B = A(\wp^{-1}(b))$ be a quadratic separable algebra with norm [B]. Then $B \cong A \times A$ if and only if $[B] \cong [o,o]$.

ii) Let $D = (a,B]$ be a quaternion algebra over A with norm form $[D] = \langle 1,-a\rangle \otimes [B]$. Let us assume the hypotesis in (1.19). Then $D \cong M_2(A)$ if and only if $[D] = [o,o] \perp [o,o]$.

§ 2. Clifford algebras.

Let (E,q) be a quadratic module over the ring A. We denote the tensor algebra of the A-module E by $T(E)$, that is $T(E) = A \oplus E \oplus \cdots \oplus E^{\otimes n} \oplus \cdots$ where $E^{\otimes n} = E \otimes \cdots \otimes E$ (n-times). In $T(E)$ we consider the ideal $J(q)$ generated by the elements $x \otimes x - q(x) \cdot 1_A$ for all $x \in E$. Then the quotient algebra

$$(2.1) \qquad C(E,q) = T(E)/J(q)$$

is called the Clifford algebra of (E,q). The inclusion $E \xrightarrow{i} T(E)$ induces a linear map $i : E \to C(E)$ and we shall denote the image of $x_1 \otimes \cdots \otimes x_n \in T(E)$ in $C(E)$ by $i(x_1)\ldots i(x_n)$. From the definition (2.1) we obtain the following relations

$$(2.2) \qquad [i(x)]^2 = q(x)$$

$$(2.3) \qquad i(x)i(y) + i(y)i(x) = b_q(x,y)$$

for all $x,y \in E$. The Clifford algebra $C(E)$ of (E,q) can be characterized by the following universal property: for every A-algebra C and every linear map $\lambda : E \to C$ with $\lambda(x)^2 = q(x)\, 1_C$ there exists a unique homomorphism of algebras $\overline{\lambda} : C(E) \to C$ such that $\lambda = \overline{\lambda} \circ i$. This follows immediately from the definition and the fact, that $C(E)$ is generated by the elements of $i(E)$. Let us, for example, consider the linear map $\alpha : E \to C(E)$, defined by $\alpha(x) = -i(x)$ for all $x \in E$. From the relation $\alpha(x)^2 = q(x)$ we obtain a homomorphism of algebras $\overline{\alpha} : C(E) \to C(E)$ with $\alpha = \overline{\alpha} \circ i$. This last relation implies $\overline{\alpha}^2 = id_{C(E)}$, that is $\overline{\alpha}$ is an isomorphism, which is called the canonical involution of $C(E)$ (we shall write simply α instead of $\overline{\alpha}$). Another important map is the canonical antiinvolution of $C(E)$, which is defined as follows: let $C(E)^o$ be the opposite algebra of $C(E)$ and define $\beta : E \to C(E)^o$ by $\beta(x) = i(x)$. Since $\beta(x)^2 = q(x)$ for all $x \in E$, there exists a homomorphism $\overline{\beta} : C(E) \to C(E)^o$ with $\beta = \overline{\beta} \circ i$. Hence $\overline{\beta}$ is an antiautomorphism of $C(E)$ with $\overline{\beta}^2 = id_{C(E)}$. We shall denote it by β. For example for $i(x_1)\ldots i(x_n) \in C(E)$ holds $\beta[i(x_1)\ldots i(x_n)] = i(x_n)\ldots i(x_1)$. The

$\mathbb{Z}/2\mathbb{Z}$-graduation $T(E)^+ = \bigoplus_{o}^{\infty} E^{\otimes 2n}$ and $T(E)^- = \bigoplus_{o}^{\infty} E^{\otimes (2n+1)}$ of $T(E)$ induces on $C(E)$ a $\mathbb{Z}/2\mathbb{Z}$-graduation $C(E)^+$, $C(E)^-$. Then $C(E) = C(E)^+ \oplus C(E)^-$ with the relations $C^+C^+ \subseteq C^+$, $C^+C^- \subseteq C^-$, $C^-C^- \subseteq C^+$.

(2.4) Remarks. 1) Let (E,q) be the trivial quadratic module over A, that is $q(x) = o$ for all $x \in E$. Then it can easily be seen that $C(E,q) \cong \Lambda(E)$ = exterior algebra of E (see [Ba]).

2) Let $(E,q) = Ae \oplus Af$ with $q(e) = a$, $q(f) = b$, $(e,f) = 1$. It follows that $C(E)$ is the 4-dimensional A-algebra $A \oplus Ae \oplus Af \oplus Aef$ with $e^2 = a$, $f^2 = b$, $ef + fe = 1$, which is a generalized quaternion algebra (see § 4).

Now we are going to compare the Clifford algebras of two given quadratic forms q, $\overline{q}$ on E, whose values differ from each other by the values of a bilinear form, that is $\overline{q}(x) = q(x) + f(x,x)$ for all $x \in E$, where $f : E \times E \to A$ is a bilinear form. Then it follows (see $[Bo]_2$).

(2.5) Lemma. There exists an isomorphism of A-modules

$$C(\overline{q}) \cong C(q).$$

Before we go into the proof of this fact, let us introduce some notation. For every linear form $\lambda \in E^*$ we define an endomorphism $\overline{\lambda}$ of $T(E)$ by

$$\overline{\lambda}(1) = o \tag{2.6}$$

$$\overline{\lambda}(x \otimes y) = -x \otimes \overline{\lambda}(y) + \lambda(x)y \tag{2.7}$$

for every $x \in E$, $y \in T(E)$. Of course this two rules determine $\overline{\lambda}$ uniquely by induction. If q is a quadratic form on E, then the ideal $J(q)$ of $T(E)$ is carried into itself by $\overline{\lambda}$, that is

$$\overline{\lambda}(J(q)) \subseteq J(q). \tag{2.8}$$

Namely, for $x \in E$, $y \in T(E)$ we have from (2.6), (2.7)

$$\overline{\lambda}((x \otimes x - q(x)) \otimes y) = \overline{\lambda}(x \otimes x \otimes y - q(x)y) =$$

$$= -x \otimes \overline{\lambda}(x \otimes y) + \lambda(x) \otimes y - q(x)\overline{\lambda}(y)$$

$$= \lambda(x)x \otimes y + x \otimes x \otimes \overline{\lambda}(y) - \lambda(x)x \otimes y - q(x)\overline{\lambda}(y)$$

$$= (x \otimes x - q(x)) \otimes \overline{\lambda}(y) \in J(q)$$

Consequently $\overline{\lambda}$ induces an endomorphism

(2.9) $$\overline{\lambda} : C(q) \to C(q)$$

An easy calculation shows that

$$\overline{\lambda}(uv) = \overline{\lambda}(u)v + \alpha(u)\,\overline{\lambda}\,(v)$$

for all $u,v \in C(q)$, i.e. $\overline{\lambda}$ is a derivation of the graded algebra $C(q)$. Another easy consequence of the above relation is the fact

(2.10) $$\overline{\lambda}^2 = o.$$

Let now $f : E \times E \to A$ be a bilinear form on E. For every $x \in E$ we define $f_x : E \to A$ by $f_x(y) = f(x,y)$ for all $y \in E$. In the same way as above we associate to the form f an endomorphism $\hat{f}$ of $T(E)$ which is inductively determined by the conditions

(2.11) $$\hat{f}(1) = 1$$

(2.12) $$\hat{f}(x \otimes u) = x \otimes \hat{f}(u) + \overline{f}_x(\hat{f}(u))$$

for all $x \in E$, $u \in T(E)$. An straightforward computation shows

(2.13) $$\hat{f} \circ \overline{\lambda} = \overline{\lambda} \circ \hat{f}$$

for every $\lambda \in E^*$, and if f,g are two bilinear forms on E, then

(2.14) $$\widehat{f+g} = \hat{f} \circ \hat{g}$$

Let us prove, for example, (2.14). We need only to check that the right side of (2.14) fulfills the conditions (2.11) and (2.12) for $f+g$. Of course $\hat{f} \circ \hat{g}(1) = \hat{f}(1) = 1$. For $x \in E$, $u \in T(E)$ we have

$$\begin{aligned}\hat{f}\circ\hat{g}(x\otimes u) &= \hat{f}(x\otimes\hat{g}(u)+\overline{g}_x(\hat{g}(u))\\ &= x\otimes\hat{f}\circ\hat{g}(u)+\overline{f}_x(\hat{f}\circ\hat{g}(u))+\hat{f}\circ\overline{g}_x(\hat{g}(u))\\ &= x\otimes\hat{f}\circ\hat{g}(u)+\overline{f}_x(\hat{f}\circ\hat{g}(u))+\overline{g}_x(\hat{f}\circ\hat{g}(u))\\ &= x\otimes\hat{f}\circ\hat{g}(u)+\overline{(f+g)}_x(\hat{f}\circ\hat{g}(u))\end{aligned}$$

which proves (2.14). In particular using (2.14) in the case $f = -g$ and the relation $\hat{o} = id$ we obtain $\hat{f}\circ\widehat{(-f)} = id$, that is

(2.15) Lemma. For every bilinear form f on E $\hat{f}: T(E) \to T(E)$ is an automorphism (of modules).

Proof of (2.5). Let $\overline{q}(x) = q(x) + f(x,x)$ for all $x \in E$, where f is a bilinear form on E. According to (2.15) f induces an automorphism $\hat{f}$ of $T(E)$. We want to prove $\hat{f}(J(\overline{q})) = J(q)$. We only need to check $\hat{f}(J(\overline{q})) \subseteq J(q)$. Now for every $x \in E$, $u \in T(E)$ we have, as it can easily be seen $\hat{f}(x\otimes x\otimes u-\overline{q}(x)u) = (x\otimes x-q(x))\otimes\hat{f}(u)$. This proves our assertion. Thus $\hat{f}$ induces an isomorphism $\overline{f}: C(\overline{q}) \xrightarrow{\sim} C(q)$ (as modules). This proves the lemma (2.5).

We now use (2.5) to prove the following

(2.16) Theorem. Let (E,q) be a quadratic module over A. Then we have an isomorphism of modules

$$C(q) \xrightarrow{\sim} \Lambda(E)$$

In particular, if E is free with basis $\{e_1,\dots,e_n\}$, then $C(q)$ is also free with basis $i(e_{j_1})\dots i(e_{j_r})$ for all $1 \leq j_1 < \dots < j_r \leq n$, $1 \leq r \leq n$.

Proof. According to (1.7), chap.I, there exists a bilinear form $f: E\times E \to A$ such that $q(x) = -f(x,x)$ for all $x \in E$, i.e. $o = q(x) + f(x,x)$. We now apply (2.5) to $\overline{q} = o$ and obtain an isomorphism

$$C(q) \xrightarrow{\sim} C(\bar{q}) = \bigwedge(E)$$

(see remark (2.4),(1)). This proves the first assertion of our proposition. It can easily be seen that this isomorphism maps $i(x_1)\ldots i(x_r) \in C(E)$ to $x_1 \wedge \ldots \wedge x_r \in \bigwedge(E)$. Then the second assertion of (2.16) follows immediately from this fact, since $e_{j_1} \wedge \ldots \wedge e_{j_r}$ for $1 \leq j_1 < \ldots < j_r \leq n$, $1 \leq r \leq n$ is a basis of E.

(2.17) <u>Corollary</u>. For every quadratic module (E,q) is $i : E \to C(E)$ injective.

According to this result we shall identify E with its image i(E) in C(E), and so we write $x \in C(E)$ instead of i(x) for every $x \in E$. With this notation, if E is free with basis $\{e_1,\ldots,e_n\}$, then the elements $e_{i_1},\ldots,e_{i_r}$, $1 \leq i_1 < \ldots < i_r \leq n$, $1 \leq r \leq n$ form a basis of C(E) over A. In particular

(2.18) Corollary. If dim E = n, then dim $C(E) = 2^n$, dim $C(E)^+ =$ dim $C(E)^- = 2^{n-1}$.

Let now (E_1,q_1), (E_2,q_2) be two quadratic modules over A. We now want to calculate the Clifford algebra of $(E_1 \perp E_2,\ q_1 \perp q_2)$. To this end it is necessary to introduce the graded tensor product of $\mathbb{Z}/2\mathbb{Z}$-graded algebras over A. Thus let B_1,B_2 be two $\mathbb{Z}/2\mathbb{Z}$-graded algebras over A. The <u>graded tensor product</u> of B_1 and B_2 is the algebra $B_1 \hat{\otimes}_A B_2$, whis has $B_1 \otimes_A B_2$ as the underlying module with the following product

$$(a_1 \otimes a_2)\cdot(b_1 \otimes b_2) = (-1)^{|a_2||b_1|} a_1b_1 \otimes a_2b_2$$

for all homogene elements $a_1,b_1 \in B_1$, $a_2,b_2 \in B_2$, where $|a|$ denotes the degree of the homogeneous element a. $B_1 \hat{\otimes}_A B_2$ has the universal property of the tensor product in the category of $\mathbb{Z}/2\mathbb{Z}$-graded algebras over A. With this notation we now can state

(2.19) Theorem. For two quadratic modules (E_1, q_1) and (E_2, q_2) holds

$$C(E_1 \perp E_2) \cong C(E_1) \hat{\otimes}_A C(E_2)$$

Proof. We consider the map $\alpha : E_1 \perp E_2 \to C(E_1) \hat{\otimes}_A C(E_2)$ given by $\alpha(e_1 + e_2) = e_1 \otimes 1 + 1 \otimes e_2$ for all $e_1 \in E_1$, $e_2 \in E_2$. It follows $[\alpha(e_1 + e_2)]^2 = (q_1(e_1) + q(e_2))\ 1 \otimes 1$, so that we obtain from the universal property of the Clifford algebra a homomorphism

$$\bar{\alpha} : C(E_1 \perp E_2) \to C(E_1) \hat{\otimes}_A C(E_2)$$

with $\alpha = \bar{\alpha} \circ i$. On the other hand we have the inclusions $\beta_j : E_j \to E_1 \perp E_2$ $(j=1,2)$, which induce homomorphisms $\bar{\beta}_j : C(E_j) \to C(E_1 \perp E_2)$ with $\bar{\beta}_j(e_j) = e_j$ for all $e_j \in E_j$ $(j=1,2)$. From the relation $e_1 e_2 = -e_2 e_1$ in $C(E_1 \perp E_2)$ for all $e_1 \in E_1$, $e_2 \in E_2$ and the universal property of $\hat{\otimes}_A$ we get a homomorphism

$$\beta : C(E_1) \hat{\otimes}_A C(E_2) \to C(E_1 \perp E_2)$$

Now it is straightforward to check that $\bar{\alpha}$ and β are inverse to each other. This proves the theorem.

(2.20) Remark. We explain below, how the Clifford algebra transforms by extensions of rings. Let $\alpha : A \to B$ be a ring homomorphism and consider a quadratic module (E,q) over A. Then α induces the extended quadratic module $(E \otimes_A B,\ q \otimes B)$ over B and the canonical inclusion $E \otimes_A B \to C(E) \otimes_A B$ has the universal property of the Clifford algebra $C(E \otimes B)$, thus the induced homomorphism

$$C(E \otimes_A B) \xrightarrow{\sim} C(E) \otimes_A B \tag{2.21}$$

is an isomorphism. For example let us consider an ideal $a \subset A$, respectively a prime ideal $p \subset A$ with the corresponding ring extensions $A \to A/a$ and $A \to A_p$, respectively. Then (2.21) shows

$$C(E/aE) \cong C(E)/aC(E) \tag{2.22}$$

(2.23) $$C(E_p) \cong C(E)_p$$

We end this section with an example, which we shall use in later applications (see (3.40)). Let us consider a quadratic separable algebra B over A with the associated quadratic norm form [B] = (B,n). We want wo calculate $C(\langle a\rangle \otimes [B])$ for $a \in A^*$. We assert

(2.24) $$C(\langle a\rangle \otimes [B]) \cong (a,B]$$

This can be seen using the universal property of C. Since $(a,B] = B \oplus Be$ with $e^2 = a$, one can define the map $\langle a\rangle \otimes [B] \to (a,B]$, $t \otimes x \to xe$ (t is the basis element of $\langle a\rangle$ with $(t,t) = a$), which has the property $(xe)^2 = xexe = x\sigma(x)e^2 = a\,n(x)$ for all $x \in [B]$. Thus it induces an homomorphism $C(\langle a\rangle \otimes [B]) \to (a,B]$, which is an isomorphism, since by reduction modulo maximal ideals it induces isomorphisms. In particular we conclude $C([B]) \cong (1,B]$, which is indeed a splitting quaternion algebra over A, that is $C([B]) \sim 1$ over A.

§ 3. The structure of Clifford algebras

Most of the results of this section are based on the following

(3.1) Theorem. Let $(E,q) = (\mathbb{H}[P],q_P)$ be the hyperbolic space associated with $P \in \underline{P}(A)$. Then there exist an isomorphism of graded A-algebras

$$\Phi_P : C(\mathbb{H}[P]) \xrightarrow{\sim} \mathrm{End}_A(\bigwedge(P))$$

In particular $C(\mathbb{H}[P])$ is a graded Azumaya algebra (see [Ba], chap.IV).

Proof. We have $\mathbb{H}[P] = P \oplus P^*$. For every $u \in P$ we define $L_u : \bigwedge P \to \bigwedge P$ by setting $L_u(x) = u \wedge x$ for all $x \in P$. For every $u^* \in P^*$ let $\overline{u}^* : \bigwedge P \to \bigwedge P$ be the endomorphism (2.9) associated with u^* (we use $\bigwedge P = C(0)$, where P is endowed with the trivial quadratic form). Now we define

$$\Phi_P : \mathbb{H}[P] \to \mathrm{End}_A(\Lambda P)$$

by $\Phi_P(u+u^*) = L_u + \bar{u}^*$. Using the relation

$$(L_u \circ \bar{u}^* + \bar{u}^* \circ L_u)(x) = \langle u,u^*\rangle x$$

for all $x \in P$ we conclude

$$[\Phi_P(u+\bar{u}^*)]^2 = (L_u + \bar{u}^*)^2 = L_{u^2} + \langle u,u^*\rangle + (\bar{u}^*)^2$$

$$= (q(u) + \langle u,u^*\rangle)\, 1$$

since $(\bar{u}^*)^2 = o$ (see (2.10)). Therefore from the universal property of the Clifford algebra follows that there exists a homomorphism of algebras

(3.2) $$\Phi_P : C(\mathbb{H}[P]) \to \mathrm{End}_A(\Lambda P)$$

which is obviously compatible with the $\mathbb{Z}/2\mathbb{Z}$-graduations of $C(\mathbb{H}[P])$ and $\mathrm{End}_A(\Lambda P)$. We recall that the later is given by the submodules

$$\mathrm{End}_A(\Lambda P)^+ = \begin{pmatrix} \mathrm{End}(\Lambda P^+) & o \\ o & \mathrm{End}(\Lambda P^-) \end{pmatrix}$$

$$\mathrm{End}_A(\Lambda P)^- = \begin{pmatrix} o & \mathrm{Hom}(\Lambda P^+,\Lambda P^-) \\ \mathrm{Hom}(\Lambda P^-,\Lambda P^+) & o \end{pmatrix}$$

of $\mathrm{End}_A(\Lambda P)$. We claim that Φ_P is an isomorphism. But Φ_P is clearly compatible with ring extensions, so that for every $m \in \max(A)$ we obtain (up to the canonical isomorphisms (2.23)) the relation $(\Phi_P)_m = \Phi_{P_m}$. Therefore we may assume, without restriction of generality, that A is a local ring (see [Bo]$_1$, ch.II, § 3). Now both $C(\mathbb{H}[P])$ as $\mathrm{End}_A(\Lambda P)$ are free A-modules of the same rank. Reducing (3.2) with respect to the maximal ideal m of A we obtain $\Phi_P(m) = \Phi_{P(m)}$, and hence we can suppose that A is a field. Now in the case of a field it can easily be seen that $\Phi_{P\oplus Q} = \Phi_P \otimes \Phi_Q$ for two vector spaces P and Q over A. This fact reduces our problem to the case dim P = 1, that is $\mathbb{H}[P] = [o,o]$. Now the assertion follows from a direct computation, which we omit. Since ΛP (in the general case) is faithfully projective, we can apply theorem (4.1), ch.IV of [Ba] to deduce the last statement of the theorem.

(3.3) Corollary. Let (E,q) be a quadratic space over A. Then C(E) is a graded Azumaya-algebra over A, and in particular a separable algebra.

Proof. We have $E \perp -E \cong \mathbb{H}[E]$ (see ch.I, (4.1)). This implies in virtue of (2.19) and (3.2)

$$\begin{aligned} C(E) \hat{\otimes}_A C(-E) &\cong C(\mathbb{H}[E]) \\ &\cong \mathrm{End}_A(\Lambda E) \end{aligned}$$

Our assertion follows from theorem (4.1), ch.IV of [Ba].

(3.4) Remark. If CENTRE (C(E)) denotes the centre of C(E) as a graded algebra, then we have shown above CENTRE (C(E)) = A. In the sequel we shall denote the centre of the ungraded algebra C(E) by Z[C(E)]. In general we have only $A \subseteq Z[C(E)]$ as well $A \subseteq Z[C(E)^+]$, but it should be noted that equality does not necessarily holds (see (3.5) below). Since the subalgebra of homogeneous elements with degree o of a separable $\mathbb{Z}/2\mathbb{Z}$-graded algebra is an ungraded separable algebra (see [Ba], ch.IV, cor.(2.3)), we conclude that $C(E)^+$ is a separable algebra over A. Let us recall the following facts on separable algebras: if C is a separable algebra over A, then C is separable over its centre Z[C], Z[C] is separable over A, and Z[C] is a direct sumand of C (see [Kn-O], ch.III, (1.6), (5.5)). Applying this results to C(E) and $C(E)^+$ we obtain, that C(E) is separable over Z[C(E)], Z[C(E)] is separable over A and is a direct sumand of C(E). The same facts hold for $C(E)^+$ and $Z[C(E)^+]$. Since both C(E) and $C(E)^+$ are faithfully projective modules, it follows that both Z[C(E)] and $Z[C(E)^+]$ are faithfully projective modules, too.

(3.5) Theorem. Let (E,q) be a quadratic space over A. Then it holds

i) if rank (E) is even, then Z[C(E)] = A and C(E) is an Azumaya algebra over A

ii) if rank (E) is odd, then $Z[C(E)^+] = A$ and $C(E)^+$ is an Azumaya algebra over A.

Proof. i) Let us consider the inclusion $A \to Z[C(E)]$. Since the relations $Z[C(E_m)] = Z[C(E)_m] = Z[C(E)]_m$ hold for all $m \in \max(A)$ (see [Kn-O], ch.III,(2.1)) we obtain, by localisation of the above inclusion, the inclusions

$$A_m \to Z[C(E_m)] = Z[C(E)]_m$$

for all $m \in \max(A)$. Hence we can suppose that A is a local ring. If m is the maximal ideal of A we only need to show $A/m = Z[C(E)](m)$. But by reduction modulo m we have $Z[C(E)](m) = Z[C(E)(m)] = Z[C(E(m))]$ (see loc.cit.), that is we have to show $A/m = Z[C(E(m))]$. Thus we have reduced the problem to the field case, where our assertion is a well-known fact (see $[Bo]_2$). The assertion (ii) is proved similarly.

(3.6) Theorem. Let (E,q) be a quadratic space over A. Then

i) if rank (E) is even, then $Z[C(E)^+]$ is a quadratic separable algebra over A;
ii) if rank (E) is odd, then $Z[C(E)]$ is a quadratic separable algebra over A.

Proof. We proceed as in the proof of (3.5)
i) Let us reduce the problem to the case of a field, where the assertion is a well-known fact (see $[Bo]_2$). Reducing the inclusion $A \to Z[C(E)^+]$ with respect to a maximal ideal m, we get

$$A/m \to Z[C(E)^+](m) = Z[C(E(m))^+]$$

(see [Kn-O], ch.III, (2.1)). Since $Z[C(E(m))^+]$ is a quadratic separable algebra over A/m (see for example (4.4), (4.5), this chapter), we use (1.1) to conclude that $Z[C(E)^+]$ is a quadratic separable algebra over A. The proof of (ii) is similar.

For any quadratic space (E,q) over A we now define

$$D(E) = \{x \in C(E) \mid xy = yx \text{ for all } y \in C(E)^+\},$$

i.e. D(E) is the centralizator of $C(E)^+$ in C(E). If $p \in \mathrm{Spec}(A)$ is a prime ideal of A, then using the identifications $C(E)_p = C(E_p)$ and $C(E)_p^+ = C(E_p)^+$ we obtain

$$D(E)_p = D(E_p) \tag{3.7}$$

Using (3.7) it is now easy to see, that D(E) is a finitely generated projective A-module (see $[Bo]_1$, ch.2).

(3.8) Lemma. i) If rank (E) is even, then $D(E) = Z[C(E)^+]$.
ii) If rank (E) is odd, then $D(E) = Z[C(E)]$.

Proof. i) In any case we have the inclusion

$$Z[C(E)^+] \to D(E),$$

as it can be easily seen. We now use (3.7) to reduce our problem to the local case and then to the field case. Thus let us assume that A is a field. Take $x \in D(E)$. Then $x = x_o + x_1$ with $x_o \in C(E)^+$, $x_1 \in C(E)^-$. For every $y \in C(E)^+$ holds $x_o y + x_1 y = y x_o + y x_1$, thus $x_o y = y x_o$, $x_1 y = y x_1$. Using a basis for E over A, it is easy to see that $x_1 = o$, that is $x = x_o \in C(E)^+$, which proves the assertion (i).
The assertion (ii) can be proved in the same way.

This result, together with (3.6), leads to the following

(3.9) Corollary. Let (E,q) be a quadratic space, whose rank is even or odd over A. Then D(E) is a quadratic separable algebra over A and a direct sumand of C(E). If rank (E) is odd, then the canonical involution of D(E) is induced by α, and if rank (E) $\equiv$ 2 (mod 4) this involution is induced by β.

The last two assertions follow from the uniqueness of the canonical involution of D(E) and the fact, that both, α in case rank (E) $\equiv$ o (mod 2) and β in case rank (E) $\equiv$ 2 (mod 4) induce non trivial involutions on D(E) (see (4.6)).
On this occasion I would like to thank Prof. M.Kneser for his remark that β in case rank (E) $\equiv$ o (mod 4) does not induce the canonical involution on D(E) (see (4.6)).

(3.10) Remark. For a quadratic space (E,q) over A holds

$$A = \{z \in C(E) \mid \alpha(z)x = xz \text{ for all } x \in E\}$$

Proof. Since the module on the right side transforms well by localisations, we may assume that A is a local ring. Hence everything is free over A. Take $z \in C(E)$ with $\alpha(z)x = xz$ for all $x \in E$. Putting $z = z_0 + z_1$ with $z_0 \in C(E)^+$, $z_1 \in C(E)^-$ we deduce $z_0 x = xz_0$ and $-z_1 x = xz_1$ for all $x \in E$. It follows $z_0 \in C(E)^+ \cap Z[C(E)] = A$ and $z_1 \in C(E)^- \cap D(E)$. Since $D(E) \subseteq C(E)^+$ if rank(E) is even, we obtain $z_1 = o$, i.e. $z = z_0 \in A$ in this case. Let now rank (E) be odd (and hence $2 \in A^*$). In this case we have $D(E) = Z[C(E)]$ (see (3.8)). Thus $z_1 \in Z[C(E)]$, that is $z_1 x = xz_1$ for all $x \in E$, which implies $z_1 x = o$, since $2 \in A^*$. Using an orthogonal basis of E we can write
$z_1 = \sum_{r \equiv 1(2)} \sum_{i_1 < \ldots < i_r} \lambda_{i_1 \ldots i_r} e_{i_1} \ldots e_{i_r}$ with $\lambda_{i_1 \ldots i_r} \in A$.
From $z_1 e_i = o$ for all $1 \leq i \leq n$ we deduce, after some straightforward computations that $z_1 = o$. This proves our assertion.

Let (E,q) be a quadratic space, whose rank is either even or odd. Then we call $D(E)$ the discriminant algebra of (E,q). According to (3.9) $D(E)$ is a quadratic separable algebra over A. Similarly we have seen that, if rank (E) is even, then $C(E)$ is an Azumaya algebra over A and if rank (E) is odd, then $C(E)^+$ is an Azumaya algebra over A. This leads us to the following.

Definition. i) The Arf-invariant of (E,q) is the class $[D(E)] \in \Delta(A)$

ii) The Witt-invariant of (E,q) is the class

$$[C(E)] \in Br(A) \qquad \text{if rank } (E) \text{ is even}$$

$$[C(E)^+] \in Br(A) \qquad \text{if rank } (E) \text{ is odd.}$$

We shall write $a(E)$ (or $a(q)$) for $[D(E)]$ and $w(E)$ (or $w(q)$) for the Witt-invariant of (E,q).

It is possible to compile this two invariants in only one, the so called Clifford invariant of (E,q). To this aim one define the graded Brauer-Wall group $BW(A)$ of similarity classes of graded Azumaya algebras (see [Ba], ch.IV, [L], $[Sm]_2$). We have shown that $C(E)$ is a graded Azumaya algebra over A, so that it defines an element

$$\tilde{w}(E) = [C(E)] \in BW(A)$$

called the Clifford invariant of (E,q). $\tilde{w}(E)$ contains all the information about the Arf and Witt invariant of (E,q) according to the exact sequences

$$o \to Br(A) \to BW(A) \to \Delta^g(A) \to o$$

$$o \to \Delta(A) \to \Delta^g(A) \to C(Spec(A), \mathbb{Z}/2\mathbb{Z})$$

(see (3.15) below), where $\Delta^g(A)$ denotes the group of graded quadratic separable algebras over A (see [Ba], ch.IV) and $C(Spec(A), \mathbb{Z}/2\mathbb{Z})$ is the group of continuous functions of $Spec(A)$ to $\mathbb{Z}/2\mathbb{Z}$. But we shall not follow this point of view, since it is more easy (at last for the author) to handle directly with the Arf and Witt invariants than with the Clifford invariant.

Another important invariant of a quadratic space (E,q) is its signed discriminant. This is defined as follows. We consider the projective submodule of rank one of $D(E)$

(3.11) $\qquad D_1(E) = \{z \in C(E) \mid \alpha(z)x = -xz \text{ for all } x \in E\}$

We claim $D_1(E) \subset D(E)$. To see this, take $z = z_o + z_1 \in D_1(E)$, where $z_o \in C(E)^+$, $z_1 \in C(E)^-$. Then $z_o x = -xz_o$, $z_1 x = xz_1$ for all $x \in E$, and since $C(E)^+$ is generated by evenly products of elements of E, we conclude $z_o y = yz_o$, $z_1 y = yz_1$ for all $y \in C(E)^+$, that is $zy = yz$ for all $y \in C(E)^+$, and hence $z \in D(E)$. From the definition (3.11) follows

$$\alpha(uv)x = \alpha(u)\alpha(v)x = -\alpha(u)xv = xuv$$

for all $u,v \in D_1(E)$, that is $uv \in A$ in virtue of (3.10). Therefore we can define a symmetric bilinear form on $D_1(E)$

(3.12) $\qquad \delta : D_1(E) \times D_1(E) \to A$

by $\delta(u,v) = uv$ for all $u,v \in D_1(E)$. A local computation shows that $(D_1(E),\delta)$ is non singular. The bilinear space $\delta(E) = (D_1(E),\delta)$ is called the signed discriminant of (E,q). For example if A is a field with $Ch(A) \neq 2$, then $\delta(E)$ is the bilinear space $\langle d_E \rangle$, where $d_E = (-1)^{n(n-1)/2}\det(E)$ ($n = \dim_A E$).

(3.13) Remark. If (E,q) is a quadratic space, whose rank is even, then $D_1(E) \subset D(E) = Z[C(E)^+]$. Hence $D_1(E) = \{z \in C(E) \mid zx = -xz \text{ for all } x \in E\}$. If (E,q) has odd rank, then $D_1(E) \subset D(E) = Z[C(E)]$ and the canonical involution of $D(E)$ is induced by α. In this case we assert $D(E) = A \oplus D_1(E)$ and $D_1(E) = \{z \in D(E) \mid \alpha(z) = -z\}$. To prove this, we first observe that $A \cap D_1(E) = o$, since $2 \in A^*$ and E is projective. Now for every $z \in D(E)$ we have

$$z = \frac{1}{2}(z+\alpha(z)) + \frac{1}{2}(z-\alpha(z)) = z_1 + z_2$$

with $\alpha(z_1) = z_1$, $\alpha(z_2) = -z_2$. It follows $z_1 \in A$ and $z_2 \in D_1(E)$, that is $D(E) = A \oplus D_1(E)$.

In the sequel we shall always assume that the rank of the modules to be considered is either even or odd. We shall now compute the Arf and Witt invariant of the orthogonal sum of two quadratic spaces.

(3.14) Lemma. Let (E_1,q_1), (E_2,q_2) be two quadratic spaces, where rank (E_1) or rank (E_2) is even. Then

$$D(E_1 \perp E_2) \cong D(E_1) \circ D(E_2)$$

Proof. In order to prove this isomorphism we may assume without restriction that A is a local ring. Therefore all modules to be considered are free. According to (2.19) one can identify $C(E_1 \perp E_2)$ with $C(E_1) \hat{\otimes} C(E_2)$. First we claim

$$D(E_1 \perp E_2) \subseteq D(E_1) \hat{\otimes} D(E_2)$$

To prove this, let us consider a basis $\{v_i\}$ of $C(E_2)$ and let $z = \Sigma w_i \otimes v_i$ be an element of $D(E_1 \perp E_2)$, where $w_i \in C(E_1)$ is uniquely determined by z for all i. In particular

$$(x \otimes 1)z = z(x \otimes 1)$$

for all $x \otimes 1 \in C(E_1)^+ \hat{\otimes} 1$. Since $(x \otimes 1)z = \Sigma xw_i \otimes v_i = z(x \otimes 1) = \Sigma w_i x \otimes v_i$, we obtain from the uniqueness of the components $xw_i = w_i x$ for all i and all $x \in C(E_1)^+$. Therefore $w_i \in D(E_1)$ for all i, that is $D(E_1 \perp E_2) \subseteq D(E_1) \hat{\otimes} C(E_2)$. Using now the same argument with a basis of $D(E_1)$ we deduce $D(E_1 \perp E_2) \subseteq D(E_1) \hat{\otimes} D(E_2)$. Let us now assume that rank (E_2) is even, that is $D(E_2) = Z[C(E_2)^+]$. In particular $D(E_1) \hat{\otimes} D(E_2) = D(E_1) \otimes D(E_2)$. We denote the canonical involutions of $D(E_1)$ and $D(E_2)$ by ρ_1 and ρ_2, respectively. Then for all $x \in E_1 \otimes E_2$, $z \in D(E_1) \otimes D(E_2)$ holds $zx = (\rho_1 \otimes \rho_2)(z)x$. Since for $z \in D(E_1 \perp E_2)$ we have $zx = xz$ for all $x \in E_1 \otimes E_2$, we obtain $[(\rho_1 \otimes \rho_2)(z)-z]x = o$ for all $x \in E_1 \otimes E_2$. This implies $(\rho_1 \otimes \rho_2)(z) = z$, that is

$$D(E_1 \perp E_2) \subseteq (D(E_1) \otimes D(E_2))^{\rho_1 \otimes \rho_2} = D(E_1) \circ D(E_2).$$

Comparing the ranks in the inclusion $D(E_1 \perp E_2) \subseteq D(E_1) \circ D(E_2)$ one easily sees, that the equality holds. This proves our proposition.

(3.15) <u>Remark</u>. The $\mathbb{Z}/2\mathbb{Z}$-graduation of $C(E)$ induces a $\mathbb{Z}/2\mathbb{Z}$-graduation on $D(E)$, that is $D(E)$ is a graded quadratic separable algebra. For such algebras one can introduce a product in the set of isomorphism classes in the following way: if B_1, B_2 are two $\mathbb{Z}/2\mathbb{Z}$-graded quadratic separable algebras with canonical involutions ρ_1 and ρ_2, respectively, we define

$$B_1 * B_2 = (B_1 \hat{\otimes}_A B_2)^{\rho_1 \otimes \rho_2},$$

which is again a $\mathbb{Z}/2\mathbb{Z}$-graded quadratic separable algebra over A. Then the set $\Delta^g(A)$ of isomorphism classes of such algebras with this product is an abelian group, which is connected with $\Delta(A)$ by the exact sequence $o \to \Delta(A) \to \Delta^g(A) \to C(\mathrm{Spec}(A), \mathbb{Z}/2\mathbb{Z})$ (see [Ba], chap.IV, §3). Using this graded product we can state the following

(3.16) Lemma. Let (E_1,q_1), (E_2,q_2) be two quadratic spaces over A. Then

$$D(E_1 \perp E_2) \cong D(E_1) * D(E_2)$$

This result can be proved in the same way as (3.14) thus we omit the proof.

Using (3.1) one easily sees that $D(\mathbb{H}[P]) \cong A \times A$ for all $P \in \underline{P}(A)$, thus we conclude from (3.14)

$$D(E \perp \mathbb{H}[P]) \cong D(E)$$

for all quadratic spaces (E,q). Hence D is an invariant for (E,q), which is compatible with the relation ~ (see chap.I, §4), that is we obtain a map

$$(3.17) \qquad a : W_q(A) \to \Delta(A)$$

Here $W_q(A)$ means the Witt ring of quadratic spaces over A, whose rank is even or odd. In the sequel we shall maintain this notation. Let $W_q(A)_o$ denote the subring of $W_q(A)$ defined by the quadratic spaces of even rank. In virtue of (3.14) it follows that the restriction of the map a (see (3.17)) to $W_q(A)_o$ is a group homomorphism $a : W_q(A)_o \to \Delta(A)$, that is

$$(3.18) \qquad a(x+y) = a(x)a(y)$$

for all $x,y \in W_q(A)_o$ (this formula is also true, if we only assume $x \in W_q(A)_o$).

We now want to calculate the Witt invariant of an orthogonal sum of quadratic spaces. In case of a field we refer the reader to [L] (specially th.(2.6), (2.8), ch.V in [L]). Let us consider two quadratic spaces (E_1,q_1), (E_2,q_2), where we assume rank (E_1) to be even. We define

$$(E_2', q_2') = (D_1(E_1), \delta) \otimes (E_2, q_2)$$

$$q_2'(x_1 \otimes e_2) = x_1^2 q_2(e_2)$$

for all $x_1 \in D_1(E_1)$, $e_2 \in E_2$. Then (see [Mi-V])

(3.19) Theorem. There exists an isomorphism

$$C(E_1 \perp E_2) \cong C(E_1) \otimes_A C(E_2')$$

Proof. Consider the canonical inclusion

$$\mu : E_2' = D_1(E_1) \otimes E_2 \to C(E_1) \hat{\otimes} C(E_2)$$

given by $\mu(x_1 \otimes e_2) = x_1 \otimes e_2$. Since

$$[\mu(x_1 \otimes e_2)]^2 = x_1^2 q_2(e_2) = q_2'(x_1 \otimes e_2),$$

there exists a homomorphism of algebras $: C(E_2') \to C(E_1) \hat{\otimes} C(E_2)$, whose image lies in the commutator of $C(E_1) \hat{\otimes} 1$, because for $e_1 \in E_1$, $e_2 \in E_2$, $x_1 \in D_1(E_1)$ we have (see (3.13))

$$(e_1 \otimes 1)(x_1 \otimes e_2) = e_1 x_1 \otimes e_2 = -x_1 e_1 \otimes e_2$$

$$= (x_1 \otimes e_2)(e_1 \otimes 1)$$

Therefore μ induces a homomorphism of algebras

$$1 \otimes \mu : C(E_1) \otimes C(E_2') \to C(E_1) \hat{\otimes} C(E_2),$$

which is certainly an isomorphism, as one easily sees by reduction modulo the maximal ideals of A. The theorem follows now from (2.19).

(3.20) Remark. With every quadratic space (E,q) we can associate the quadratic space $\nabla(E) = (D(E), n)$ (see § 1), where n is the norm form of $D(E)$. Then it is easy to see that $D(\nabla(E)) \cong D(E)$, $C(\nabla(E)) \sim 1$ and $(D_1(\nabla(E)), \delta) \cong (D_1(E), \delta)$. We shall call $\nabla(E)$ the discriminant form of (E,q).

Let now (E_1,q_1), (E_2,q_2) be two quadratic spaces, whose ranks are even. From (3.19) and (3.20) follows

$$C(\delta(E_1) \otimes E_2) \otimes C(\nabla(E_1)) \cong C(E_2) \hat{\otimes} C(\nabla(E_1))$$

$$\cong C(E_2) \otimes C(\delta(E_2) \otimes \nabla(E_1)).$$

On the other hand we deduce again from (3.19)

$$C(\nabla(E_1) \perp \nabla(E_2)) \cong C(\nabla(E_1)) \hat{\otimes} C(\nabla(E_2))$$

$$\cong C(\delta(E_2) \otimes \nabla(E_1)) \otimes C(\nabla(E_2))$$

$$\cong C(\nabla(E_1)) \otimes C(\delta(E_1) \otimes \nabla(E_2))$$

Combining this isomorphisms we get

(3.21) Proposition. Let (E_i,q_i), $i=1,2$ be two quadratic spaces over A whose ranks are even. Then

$$C(E_1 \perp E_2) \otimes C(\nabla(E_1)) \otimes C(\nabla(E_2))$$

$$\cong C(E_1) \otimes C(E_2) \otimes C(\nabla(E_1) \perp \nabla(E_2))$$

$$\cong C(E_1) \otimes C(E_2) \otimes C(\nabla(E_1)) \otimes C(\delta(E_1) \otimes \nabla(E_2))$$

With this result we are now able to calculate $w(E_1 \perp E_2)$ in function of $w(E_1)$ and $w(E_2)$ in $Br(A)$. As remarked before (see end of § 2), $w([B]) = 1$ for any quadratic separable algebra B over A, thus in particular $w(\nabla(E)) = 1$ in $Br(A)$. Hence for two quadratic spaces E_1, E_2 of even rank we have in $Br(A)$

(3.22) $$w(E_1 \perp E_2) = w(E_1)\, w(E_2)\, w(\delta(E_1) \otimes \nabla(E_2))$$

Next we consider the case, where both ranks are odd. Let (E_i,q_i) $i=1,2$ be two quadratic spaces over A.

(3.23) Lemma. If rank (E_1) and rank (E_2) are odd, then

$$C(\delta(E_1) \otimes \nabla(E_2)) \tilde{\to} D(E_1) \hat{\otimes} D(E_2)$$

(here we consider $D(E_1), D(E_2)$ as $\mathbb{Z}/2\mathbb{Z}$-graded algebras).

Proof. Consider the canonical inclusion $D_1(E) \otimes D(E_2) \to D(E_1) \otimes D(E_2)$ given by $x_1 \otimes x_2 \to x_1 \otimes x_2$. Since rank (E_2) is odd, thus we have $D(E_2) = A \oplus D_1(E_2)$ (see (3.13)), that is $x_2 = a + y$ with $a \in A$, $y \in D_1(E_2)$ for every $x_2 \in D(E_2)$. Therefore for any $x_1 \in D_1(E_1)$ holds (in $D(E_1) \hat{\otimes} D(E_2)$)

$$(x_1 \otimes x_2)^2 = (x_1 \otimes a + x_1 \otimes y)^2 = x_1{}^2 \otimes a^2 - x_1{}^2 \otimes y^2$$

$$= x_1{}^2 \otimes (a^2 - y^2) = x_1{}^2(a^2 - y^2)\, 1 \otimes 1$$

$$= q(x_1 \otimes x_2)\, 1 \otimes 1$$

where q is the quadratic form of $D_1(E_1) \otimes D(E_2)$. Therefore there exists a homomorphism of algebras $C(\delta(E_1) \otimes D(E_2)) \to D(E_1) \hat{\otimes} D(E_2)$, which indeed is an isomorphism

(3.24) Lemma. Let (E,q) be a quadratic space with odd rank. Then

$$C(E) \cong C(E)^+ \otimes D(E)$$

Proof. Since rank (E) is odd we have $D(E) = Z[C(E)]$. The inclusion $C(E)^+ \to C(E)$ induces a homomorphism of algebras

$$C(E)^+ \otimes D(E) \to C(E)$$

given by $x \otimes z \to xz$ for $x \in C(E)^+$, $z \in D(E)$. Because both modules have the same rank and the homomorphism above is compatible with reductions modulo maximal ideals, we only need to prove our assertion in the field case, in which case this is a well-known fact (see $[Bo]_2$, § 9, Nr.4). It should be noted that this isomorphism is a graded one, where the graduation of $C(E)^+$ is concentrated in o and the graduation of $D(E)$ is induced by $C(E)$.

Let us now consider two quadratic spaces (E_1, q_1), (E_2, q_2) whose ranks are both odd. From (2.19) and (3.24) we get

$$C(E_1 \perp E_2) \cong C(E_1) \hat{\otimes} C(E_2)$$

$$\cong C(E_1)^+ \otimes D(E_1) \hat{\otimes} C(E_2)^+ \otimes D(E_2)$$

$$\cong C(E_1)^+ \otimes C(E_2)^+ \otimes D(E_1) \hat{\otimes} D(E_2)$$

Consequently, from (3.23) we may conclude

(3.25) Theorem. For two quadratic spaces (E_1, q_1), (E_2, q_2), whose ranks are both odd it holds

$$C(E_1 \perp E_2) \cong (C(E_1)^+ \otimes C(E_2)^+ \otimes C(\delta(E_1) \otimes \nabla(E_2))$$

In particular in Br(A) we obtain

(3.26) $$w(E_1 \perp E_2) = w(E_1)\, w(E_2)\, w(\delta(E_1) \otimes \nabla(E_2))$$

We want now to find a formula for $w(E_1 \perp E_2)$, where rank (E_1) is odd (but we do not impose any restriction on rank (E_2)). To this end let us consider the subalgebra

(3.27) $$B = 1 \hat{\otimes} C(E_2)^+ + D_1(E_1) \hat{\otimes} C(E_2)^-$$

of $C(E_1 \perp E_2)^+ \subset C(E_1) \hat{\otimes} C(E_2)$. Then it holds

(3.28) Lemma. There exists an isomorphism of algebras (ungraded)

$$C(-\delta(E_1) \otimes E_2) \cong B$$

Proof. Consider the inclusion $D_1(E_1) \otimes E_2 \to B$. Then for all $x \in D_1(E_1)$, $y \in E_2$, it follows that

$$(x \otimes y)^2 = -x^2 \otimes y^2 = -x^2 q_2(y)\, 1 \otimes 1,$$

thus it induces a homomorphism of algebras

$$C(-\delta(E_1) \otimes E_2) \to B$$

which is obviously compatible with reductions modulo maximal ideals. Thus to prove that this map is an isomorphism we may assume that A is a field, in which case the proof is straightforward.

(3.29) Theorem. Let (E_1, q_1) be a quadratic space, whose rank is odd. Then for every quadratic space (E_2, q_2) it holds

$$C(E_1 \perp E_2)^+ \cong C(E_1)^+ \otimes C(-\delta(E_1) \otimes E_2)$$

Proof. Since the subalgebras $C(E_1)^+ \otimes 1$ and B of $C(E_1 \perp E_2)^+$ commute, we obtain an homomorphism

$$C(E_1)^+ \otimes B \to C(E_1 \perp E_2)^+$$

given by $x \otimes y \to (x \otimes 1)y$ for all $x \in C(E_1)^+$, $y \in B$. Again using reduction modulo maximal ideals of A one sees that this map is an isomorphism.

In particular we obtain in Br(A)

$$(3.30) \qquad w(E_1 \perp E_2) = w(E_1)\, w(-\delta(E_1) \otimes E_2)$$

if rank (E_1) is odd and rank (E_2) is even. We shall now write this formula in a more explicite form. Since rank (E_1) is odd, we have $\delta(-E_1) \cong -\delta(E_1)$, and therefore the formula (3.29) reads $C(E_1 \perp E_2)^+ \cong C(E_1)^+ \otimes C(\delta(-E_1) \otimes E_2)$. We now tensor this relation with $C(\nabla(-E_1))$ and use (3.19) to conclude

$$C(E_1 \perp E_2)^+ \otimes C(\nabla(-E_1)) \cong C(E_1)^+ \otimes C(\nabla(-E_1)) \hat{\otimes} C(E_2),$$

since $\delta(\nabla(E)) = \delta(E)$ for any space (E,q). We use again (3.19) and the relation $C(\delta(E) \otimes \nabla(F)) \cong C(\delta(F) \otimes \nabla(E))$, which is valid for any two spaces E,F, and we get

$$C(E_1 \perp E_2)^+ \otimes C(\nabla(-E_1)) \cong C(E_1)^+ \otimes C(E_2) \otimes C(\delta(-E_1) \otimes \nabla(E_2)).$$

Hence the formula (3.30) reads now

(3.31) $$w(E_1 \perp E_2) = w(E_1)\, w(E_2)\, w(-\delta(E_1) \otimes \nabla(E_2))$$

One advantage of the Clifford invariant $\widetilde{w}$ is the fact, that all the above formulas can be compiled in only one, namely

(3.32) $$\widetilde{w}(E_1 \perp E_2) = \widetilde{w}(E_1)\, \widetilde{w}(E_2)$$

in BW(A), where E_1, E_2 are any two quadratic spaces over A (see (2.19)).

Now we shall study, shortly, the invariants of the quadratic space $\langle a\rangle \otimes E$ for any mit $a \in A^*$ and any quadratic space (E,q). Let us begin with an space E of odd rank. From (3.29) it follows

(3.33) Proposition. For any $a \in A^*$ and any space E of odd rank it holds

$$C(\langle a\rangle \otimes E)^+ \otimes C(\mathbb{H}) \cong C(E)^+ \otimes C(\mathbb{H})$$

Proof. Let us apply (3.29) to the quadratic space $[-a] \perp F$, where F is any quadratic space (note that $2 \in A^*$). Then $C([-a] \perp F)^+ \cong C([-a])^+ \otimes C(-\delta[-a] \otimes F)$. Since $\delta[-a] = \langle -a\rangle$ and $C([-a])^+ \cong A$ we conclude

(3.34) $$C([-a] \perp F)^+ \cong C(\langle a\rangle \otimes F)$$

Putting $F = [a] \perp E$ in (3.34) (note that $\langle a\rangle \otimes F = \langle a\rangle \otimes E \perp [1]$) we obtain

$$C(\langle a\rangle \otimes E \perp [1]) \cong C([-a] \perp [a] \perp E)^+ \cong C(\mathbb{H} \perp E)^+$$

Now according to (3.25) we have $C(\langle a\rangle \otimes E \perp [1]) \cong C(\langle a\rangle \otimes E)^+ \otimes C([1])^+ \otimes C(\delta(\langle a\rangle \otimes E) \otimes \nabla[1]) \cong C(\langle a\rangle \otimes E)^+ \otimes C(\mathbb{H})$ because $\nabla([1]) \cong \mathbb{H}$. On the other hand it holds $C(E \perp \mathbb{H})^+ \cong C(E)^+ \otimes C(\mathbb{H})$. Hence, inserting these two results in the above isomorphism, we conclude $C(\langle a\rangle \otimes E)^+ \otimes C(\mathbb{H}) \cong C(E)^+ \otimes C(\mathbb{H})$.

In particular this shows that

(3.35) $$w(\langle a\rangle \otimes E) = w(E)$$

for any space E of odd rank. The Arf-invariant of $\langle a\rangle \otimes E$ in this case is given by

$$D(\langle a\rangle \otimes E) \cong D(E) \circ A(\sqrt{-a}) \tag{3.36}$$

($A(\sqrt{-a})$ is a quadratic separable algebra with involution $\sqrt{-a} \to -\sqrt{-a}$, because $2 \in A^*$). Using (3.29) and (3.31) for a space E_1 of odd rank we deduce $C(E_1)^+ \otimes C(-\delta(E_1) \otimes E_2) \otimes C(\nabla(-E_1)) \cong$ $C(E_1)^+ \otimes C(E_2) \otimes C(-\delta(E_1) \otimes \nabla(E_2))$. Now taking $E_1 = [-a]$, $a \in A^*$, and $E_2 = E$ we get

$$C(\langle a\rangle \otimes E) \otimes C(\nabla[a]) \cong C(E) \otimes C(\langle a\rangle \otimes \nabla(E)). \tag{3.37}$$

Thus if rank (E) is even, it follows that

$$w(\langle a\rangle \otimes E) = w(E)\, w(\langle a\rangle \otimes \nabla(E)). \tag{3.38}$$

It should be noted that the above proof of (3.38) is valid under the assumption $2 \in A^*$, since we have used the quadratic space $[a]$. However the formula (3.38) is true for any commutative ring A, but we shall omit the details of the proof in the general case. For the Arf invariant of $\langle a\rangle \otimes E$, where rank (E) is even, we obtain

$$a(\langle a\rangle \otimes E) = a(E) \tag{3.39}$$

More generally, if $b = \langle a_1, \ldots, a_n\rangle$ is a bilinear space over A, $a_i \in A^*$, and $d(b) = (-1)^{n(n-1)/2} a_1 \ldots a_n$ is the signed determinant of b, then the following formulas hold for any quadratic space E

$$a(b \otimes E) = \begin{cases} a(E)^n & \text{if rank (E) is even} \\ a(E)^n [A(\sqrt{d(b)})] & \text{if rank (E) is odd} \end{cases} \tag{3.40}$$

$$w(b \otimes E) = \begin{cases} w(E)^n\, w(\langle d(b)\rangle \otimes \nabla(E)) & \text{if rank (E) is even} \\ w(E)^n & \text{if rank (E) is odd.} \end{cases} \tag{3.41}$$

(3.42) Remark. Let us consider a quadratic separable algebra B over the ring A with its associated quadratic norm form $[B]$. We have proved $C([B]) \cong (1,B]$ and $D([B]) \cong B$, that is $w([B]) = 1$ and $a([B]) = [B] \in \Delta(A)$. More generally if $a \in A^*$ then it holds $w(\langle a\rangle \otimes [B]) = [(a,B]] \in Br(A)$ and $a(\langle a\rangle \otimes [B]) = [B] \in \Delta(A)$. In particular if B_1, B_2 are two such algebras with $[B_1] \cong [B_2]$ (norm forms), then $B_1 \cong D([B_1]) = D([B_2]) \cong B_2$, which proves the lemma (1.18). Let us now consider a quaternion algebra $Q = (a,B]$ with its associated norm form $[Q] = [B] \perp \langle -a \rangle \otimes [B]$. Now (3.19) implies $C([Q]) \cong C([B]) \hat{\otimes} C(\langle -a \rangle \otimes [B]) \cong C([B]) \otimes C(\langle -a \rangle \otimes \delta([B]) \otimes [B])$. We claim $\delta([B]) \otimes [B] \cong -[B]$, where the isomorphism is given by $x \otimes y \to xy$ for all $x \in D_1([B])$, $y \in [B]$, here the product xy is to be taken in $C([B])$. This is well-defined, since $D_1([B]) \subseteq C([B])^+$ and $[B] = C([B])^-$. Furthermore we have in $[B]$ $q(xy) = (xy)^2 = xyxy = -x^2y^2$ (see (3.13)), that is $q(xy) = -\delta(x,x)q(y)$, which proves the claim. Using this isomorphism it follows that

$$C([Q]) \cong C([B]) \otimes C(\langle a \rangle \otimes [B])$$

$$C([Q]) \cong C([B]) \otimes Q$$

In particular we obtain $w([Q]) = [Q] \in Br(A)$. From the above isomorphism one easily deduces the relation $D([Q]) \cong B \circ B \cong A \times A$, that is $a([Q]) = 1$. Using this remarks it is now easy to finish the proof of lemma (1.19).

(3.43) <u>Summary</u>. We sum up the most important formulas, which enable to compute the Arf and Witt invariants of orthogonal sums of quadratic spaces. Let E,F be two quadratic spaces whose ranks can be either even or odd. Then we have proved

i) $a(E \perp F) = a(E)a(F)$ if rank $(E)\cdot$rank (F) is even.

ii) $w(E \perp F) = w(E)w(F)w(\delta(E) \otimes \nabla(F))$ if rank (E) and rank (F) are both even or odd.

iii) $w(E \perp F) = w(E)w(F)w(-\delta(E) \otimes \nabla(F))$ if rank (E) is odd and rank(F) is even.

iv) $w(\langle a \rangle \otimes E) = w(E)$ if rank(E) is odd.

v) $w(\langle a \rangle \otimes E) = w(E)\, w(\langle a \rangle \otimes \nabla(E))$ if rank(E) is even.

vi) $a(\langle a \rangle \otimes E) = a(E)[A(\sqrt{-a})]$ if rank(E) is odd.

vii) $a(\langle a \rangle \otimes E) = a(E)$ if rank(E) is even.

§ 4. Some computations.

The purpose of this section is to compute the Arf invariant of quadratic spaces, which have a simple orthogonal decomposition. Let us first consider spaces of lower dimension.

(4.1) Let $(E,q) = (Ae,q)$ with $q(e) = a \in A^*$. If $2 \in A^*$ then E is non singular. In any case we have $C(E) = A \oplus Ae$ with $e^2 = a$, that is $C(E) = A(\sqrt{a})$. In particular $D(E) = A(\sqrt{a})$, $\delta(E) = (Ae,\delta)$ with $\delta(e,e) = a$, i.e. $\delta(E) = \langle a\rangle$.

(4,2) Let $(E,q) = [a,b]$ with $1 - 4ab \in A^*$, where $E = Ae \oplus Af$, $q(e) = a$, $q(f) = b$, $(e,f) = 1$. Then $C(E) = A \oplus Ae \oplus Af \oplus Aef$, $e^2 = a$, $f^2 = b$, $ef + fe = 1$. In particular $D(E) = A \oplus Aef$ with $(ef)^2 = ef - ab$, that is $D(E) = A(\wp^{-1}(-ab))$. We have also $\delta(E) = (A(1 - 2ef),\delta) = \langle 1 - 4ab\rangle$. If we assume $a \in A^*$, then $[a,b] \cong \langle a\rangle \otimes [1,ab]$ and hence $C(E) = (a,-ab]$.

(4.3) Let $(E,q) = [c] \perp [d]$ with $c,d \in A^*$. Then $C(E) = A \oplus Ag \oplus Ah \oplus Agh$ with $g^2 = c$, $h^2 = d$, $gh = -hg$. Then $D(E) = A \oplus Agh$ if $2 \in A^*$. In this case we have $D(E) = A(\sqrt{-cd})$. Moreover $E = [c] \perp [d] \cong [c, 1 + 4cd/4c] = \langle c\rangle \otimes [1, 1 + 4cd/4]$. Hence $C(E) \cong (c, -\frac{1+4cd}{4}]$.

(4.4) Proposition. Assume $E = [a_1,b_1] \perp \ldots \perp [a_n,b_n]$ with basis $\{e_1,f_1,\ldots,e_n,f_n\}$, where $q(e_i) = a_i$, $q(f_i) = b_i$, $(e_i,f_i) = 1$ $(1 - 4a_ib_i \in A^*)$ for all i. Let us define $z_I = \prod_{i\in I} e_if_i \in C(E)$ for every subset $I \subset \{1,\ldots,n\}$. Then we have $D(E) = A \oplus Az$ where

$$z = \sum_{I\neq\emptyset} (-2)^{|I|-1} z_I$$

If $d_I = \prod_{i\in I} (-a_ib_i)$ and $d = \sum_{I\neq\emptyset} 4^{|I|-1} d_I \in A$, then

$$z^2 = z + d, \qquad \text{that is } D(E) = A(\wp^{-1}(d)).$$

Proof. We make induction on n. The case $n = 1$ was treated in (4.2).

Assume now $n > 1$ and set $E_1 = [a_1, b_1] \perp \dots \perp [a_{n-1}, b_{n-1}]$, $E_2 = [a_n, b_n]$. Then from (3.10) it follows that $D(E) = D(E_1) \circ D(E_2)$, where $D(E_1) = A \oplus Az_1$, $D(E_2) = A \oplus Az_2$, $z_2 = e_n f_n$ and by induction

$$z_1 = \sum_{\substack{J \subset \{1,\dots,n-1\} \\ J \neq \emptyset}} (-2)^{|J|-1} z_J$$

Since $D(E) = A \oplus Az$ with $z = z_1 + z_2 - 2z_1 z_2$, $z^2 = z + d$, $d = d_1 + d_2 + 4d_1 d_2$, where $z_1^2 = z_1 + d_1$, $z_2^2 = z_2 + d_2$, we obtain

$$z = \sum_{\substack{J \subset \{1,\dots,n-1\} \\ J \neq \emptyset}} (-2)^{|J|-1} z_J + e_n f_n - 2 \sum_{\substack{J \subset \{1,\dots,n-1\} \\ J \neq \emptyset}} (-2)^{|J|-1} z_J e_n f_n$$

$$z = \sum_{\substack{J \subset \{1,\dots,n-1\} \\ J \neq \emptyset}} (-2)^{|J|-1} z_J + z_{\{n\}} + \sum_{\substack{J \subset \{1,\dots,n-1\} \\ J \neq \emptyset}} (-2)^{|J \cup \{n\}|-1} z_{J \cup \{n\}}$$

$$= \sum_{\substack{I \subset \{1,\dots,n\} \\ I \neq \emptyset}} (-2)^{|I|-1} z_I$$

The expression for d follows similarly. This proves our assertion.

If $4 = o$ in A, we obtain from (4.4) $D(E) = A \oplus Az$ with $z^2 = z + d$, $d = \sum_{i=1}^{n} a_i b_i$, that is

$$D(E) = A(\wp^{-1}(\sum_{i=1}^{n} a_i b_i)).$$

If we identify $\Delta(A)$ with $A/\wp(A)$ (see (1.11)), then the Arf invariant of $E = [a_1, b_1] \perp \dots \perp [a_n, b_n]$ is given by the class of $a_1 b_1 + \dots + a_n b_n$ in $A/\wp(A)$. Thus for a field of characteristic 2 we obtain the classical Arf invariant (see [A]).

(4.5) Proposition. Let (E,q) be a quadratic space with an orthogonal decomposition $E = [c_1] \perp \dots \perp [c_m]$ where $c_i \in A^*$ for all i (in particular $2 \in A^*$). If $\{y_1, \dots, y_m\}$ is the corresponding orthogonal basis with $q(y_i) = c_i$, then $D(E) = A \oplus Ay$, where

$$y = y_1 \dots y_m \quad \text{and} \quad y^2 = (-1)^{\frac{m(m-1)}{2}} c_1 \dots c_m,$$

that is $D(E) = A(\sqrt{(-1)^{m(m-1)/2} c_1 \dots c_m})$ (see (3.40)).

Proof. Using (3.16) we can apply induction to prove the proposition. We omit the details. It should be noted that the canonical involution of $D(E)$ is given by $y \to -y$.

(4.6) Example. Let us consider a field A and a quadratic space (E,q) with rank $(E) = 2n$.

i) If $\mathrm{Ch}(A) \neq 2$, then $E = \langle y_1 \rangle \perp \dots \perp \langle y_{2n} \rangle$ with $q(y_i) = c_i \neq 0$. From (4.5) we conclude $D(E) = A \oplus Ay$ with $y = y_1 \dots y_{2n}$. The involution β (of $C(E)$) acts on y by $\beta(y) = (-1)^n y$.

ii) If $\mathrm{ch}(A) = 2$, then $E = \langle e_1, f_1 \rangle \perp \dots \perp \langle e_n, f_n \rangle$ with $(e_i, f_i) = 1$. Hence (see (4.4)) $D(E) = A \oplus Az$ with $z = e_1 f_1 + \dots + e_n f_n$, and therefore $\beta(z) = f_1 e_1 + \dots + f_n e_n = n + z$.

We conclude that in both cases β induces the canonical involution of $D(E)$ if and only if $n \equiv 1 \pmod 2$, i.e. $\dim E \equiv 2 \pmod 4$ (compare (3.9)). If $4 \mid \dim E$ then β induces the identity on $D(E)$.

§ 5. Quadratic spaces of lower dimension.

In this section we shall assume throughout that the ring A has the following properties:

i) In Quad(A) holds the cancelation law

ii) Every Azumaya algebra over A is determined (up to isomorphism) by its rank and class in $\mathrm{Br}(A)$. These properties are fulfilled for example for semilocal rings (see ch.III, (4.3) and [Kn]). Let us recall the following well-known facts over fields: if q_1, q_2 are two quadratic spaces with $\dim q_1 = \dim q_2 \leq 3$, $a(q_1) = a(q_2)$,

$w(q_1) = w(q_2)$, then $q_1 \cong q_2$ (see [A], [W]); if $\dim q_1 = \dim q_2 = 4$, $a(q_1) = a(q_2)$, $w(q_1) = w(q_2)$ and q_1, q_2 represent a common element, then $q_1 \cong q_2$ too. We now want to check these results for the rings we are considering. First, in the two-dimensional case we have

(5.1) Proposition. Let $E_1 = \langle a_i\rangle \otimes [B_i]$ be two quadratic spaces of rank 2 with $a(E_1) = a(E_2)$ and $w(E_1) = w(E_2)$. Then $E_1 \cong E_2$.

Proof. Since $D(E_1) \cong D(E_2)$, it follows that $B_1 \cong B_2$, and hence $[B_1] \cong [B_2]$. On the other hand, $w(E_1) = w(E_2)$ together with the assumption (ii) imply $C(E_1) = C(E_2)$. Therefore (see (3.42))

$$[B_1] \perp \langle -a_1\rangle \otimes [B_1] \cong [B_2] \perp \langle -a_2\rangle \otimes [B_2]$$

and in virtue of the assumption (i) it follows that
$\langle a_1\rangle \otimes [B_1] \cong \langle a_2\rangle \otimes [B_2]$.

Let us now consider spaces of dimension 3. We shall treat only the very simple case where the spaces have an orthogonal basis, that is we assume $E_i = \langle a_i\rangle \perp \langle b_i\rangle \perp \langle c_i\rangle$, $(i=1,2)$ $a_i, b_i, c_i \in A^*$. If $a(E_1) = a(E_2)$, then $a_1b_1c_1 \equiv a_2b_2c_2 \pmod{A^{*2}}$. Now, scaling both spaces with $\langle a_1b_1c_1\rangle \cong \langle a_2b_2c_2\rangle$ we may assume
$E_i = \langle a_i\rangle \perp \langle b_i\rangle \perp \langle a_ib_i\rangle$. Since we are assuming $w(E_1) = w(E_2)$, i.e. $C(E_1)^+ \cong C(E_2)^+$, we conclude $C(\langle 1\rangle \perp E_1) = C(\langle 1\rangle \perp E_2)$ (see (3.25)). Now, one easily shows that there exist two quaternion algebras Q_1, Q_2 such that $\langle 1\rangle \perp E_i \cong [Q_i]$ $(i=1,2)$. Again we use the assumption (ii) and (3.42) to conclude that $[Q_1] \cong [Q_2]$, that is $\langle 1\rangle \perp E_1 \cong \langle 1\rangle \perp E_2$. Thus $E_1 \cong E_2$ (use (i)). We have proved the following.

(5.2) Proposition. Let $E_i = \langle a_i\rangle \perp \langle b_i\rangle \perp \langle c_i\rangle$ $(i=1,2)$ be two quadratic spaces with $a(E_1) = a(E_2)$, $w(E_1) = w(E_2)$. Then $E_1 \cong E_2$.

As a matter of exercise we shall now treat the case $\dim E_1 = \dim E_2 = 4$ over a ring A with $4 = o$. Thus let be given two quadratic spaces

$E_i = \langle a_i\rangle \otimes [1,b_i] \perp \langle c_i\rangle \otimes [1,d_i]$ with $a_1 = a_2$, $a(E_1) = a(E_2)$, $w(E_1) = w(E_2)$. Scaling with $a_1 = a_2$ both spaces, we may assume $a_1 = a_2 = 1$. Using (3.19) and (4.2) we conclude $w(E_1) = [(c_1,-d_1]] = w(E_2) = [(c_2,-d_2]]$ (here we have used $4 = o$), that is $(c_1,-d_1] \cong (c_2,-d_2]$ (see (ii)). In particular the norm forms of these algebras are isomorphic, that is $\langle 1,-c_1\rangle \otimes [1,d_1] \cong \langle 1,-c_2\rangle \otimes [1,d_2]$. From the assumption (i) it follows that $[1,d_1] \perp \langle c_2\rangle \otimes [1,d_2] \cong [1,d_2] \perp \langle c_1\rangle \otimes [1,d_1]$. Let us define the spaces $F_1 = E_1 \perp [1,d_1] \perp [1,d_2]$ and $F_2 = E_2 \perp [1,d_1] \perp [1,d_2]$. In virtue of (i) we only need to show $F_1 \cong F_2$. But from the above isomorphism and (i) this is equivalent with (*): $[1,b_1] \perp [1,d_1] \cong [1,b_2] \perp [1,d_2]$. It should be noted that the Arf and Witt invariants of these spaces are equal. Now, using (1.10), (3.14) and $a(E_1) = a(E_2)$ we conclude that $A(\wp^{-1}(-b_1-d_1)) \cong A(\wp^{-1}(-b_2-d_2))$, i.e. $[1,b_1+d_1] \cong [1,b_2+d_2]$. But on the other hand holds $[1,b_1] \perp [1,d_1] \cong [1,b_1+d_1] \perp E_o$, $[1,b_2] \perp [1,d_2] \cong [1,b_2+d_2] \perp F_o$ with $\dim E_o = \dim F_o = 2$, $a(E_o) = a(F_o)$, $w(E_o) = w(F_o)$. From (5.1) it follows that $E_o \cong F_o$, which proves (*). In particular we get $E_1 \cong E_2$. Thus we have proved.

(5.3) <u>Proposition</u>. Let A be a ring with $4 = o$, (i) and (ii). If $E_i = \langle a\rangle \otimes [1,b_i] \perp \langle c_i\rangle \otimes [1,d_i]$ are two spaces with $a(E_1) = a(E_2)$, $w(E_1) = w(E_2)$, then $E_1 \cong E_2$.

CHAPTER III

The orthogonal group

§ 1 Notations.

Let (E,q) be a quadratic space over the ring A. An automorphism σ of (E,q) is a linear isomorphism $\sigma : E \xrightarrow{\sim} E$ such that $q(\sigma(x)) = q(x)$ for all $x \in E$. We shall denote the group of all automorphisms of (E,q) by $O(E,q)$ (or simply $O(E)$ or $O(q)$). This group is called the orthogonal group of (E,q). If $A \to B$ is a ring homomorphism, then the scalar extension $E \to E \otimes B$ induces a group homomorphism $O(E) \to O(E \otimes B)$. In particular every ideal $a \subset A$ induces the canonical reduction homomorphism

$$\varphi_a : O(E) \to O(E(a)), \tag{1.1}$$

whose kernel will be denoted by $O(E,a)$. If $A = A_1 \times \ldots \times A_a$ is a direct product of rings, then the quadratic space (E,q) has a decomposition $(E,q) = (E_1,q_1) \times \ldots \times (E_s,q_s)$, where (E_i,q_i) is a quadratic space over A_i. In this case we obtain a decomposition for $O(E)$, namely

$$O(E,q) = O(E_1,q_1) \times \ldots \times O(E_s,q_s) \tag{1.2}$$

In particular this fact can be applied to semi local rings. If A is a semi local ring with maximal ideals $m_1,\ldots,m_s$ and $r = m_1 \cap \ldots \cap m_s$ is the Jacobson radical of A, then $A/r = A/m_1 \times \ldots \times A/m_s$. Hence we obtain $(E(r),q(r)) = \prod_m (E(m),q(m))$, where $(E(m),q(m))$ is the reduction of (E,q) with respect to the maximal ideal m. Therefore

$$O(E(r)) = \prod_m O(E(m)) \tag{1.3}$$

In this chapter we shall study the orthogonal group $O(E)$ of a quadratic space (E,q) over a semi local ring A. The reduction homomorphism $\varphi_r : O(E) \to O(E(r))$ plays an important role on this subject, because it turns out to be onto. As a consequence of this fact we shall deduce the cancellation law for quadratic spaces over semi local rings (see $[K]_2$, $[Kne]_2$, [R]).

The following automorphisms of the quadratic space (E,q) shall be used continuously in this chapter:

(1.4) Hyperplane reflections. Let $x \in E$ be a strictly anisotropic element, that is $q(x) \in A^*$. We define a map

$$\sigma_x : E \to E$$

by
$$\sigma_x(y) = y - \frac{(y,x)}{q(x)} x$$

for all $y \in E$ (we have written (y,x) instead of $b_q(y,x)$, and we shall maintain this notation throughout this chapter). The map σ_x is called a hyperplane reflection (or a symmetry) of E. σ_x is evidently linear and it has the properties $\sigma_x \circ \sigma_x = id_E$, $q(\sigma_x(y)) = q(y)$ for all $y \in E$, that is $\sigma_x \in O(E)$. From the definition one sees that σ_x leaves $\langle x\rangle^{\perp}$ elementwise fixed and maps x to -x. The subgroup of O(E), which is generated by all hyperplane reflections will be denoted by S(E).

(1.5) Siegel-Transvection. Consider a pair of elements $x,y \in E$ such that $q(x) = o$, $y \in \langle x\rangle^{\perp}$ (that is $(x,y) = o$). Then we define the linear map

$$E(x,y) : E \to E$$

by $E(x,y)(z) = z + (z,x)y - (z,y)x - q(y)(z,x)x$

for all $z \in E$. An easy computation shows that $q(E(x,y)(z)) = q(z)$ holds for all $z \in E$. If $y' \in \langle x\rangle^{\perp}$ is another element orthogonal to x we get

$$(1.6) \qquad E(x,y+y') = E(x,y) \circ E(x,y')$$

and in particular $E(x,y) \circ E(x,-y) = id_E$. Hence $E(x,y) \in O(E)$. For example if $y \in \langle x\rangle^{\perp}$ is strictly anisotropic, then

$$(1.7) \qquad E(x,y) = \sigma_{y-q(y)x} \circ \sigma_y$$

The automorphism E(x,y) is called Siegel transvection. The subgroup of O(E) generated by all Siegel transvections will be denoted by $\hat{\mathbb{E}}(E)$.

(1.8) Remark. S(E) and $\hat{\mathbb{E}}(E)$ are normal subgroups of O(E). This follows immediately from the following relations: for all $\sigma \in O(E)$, $z \in E$ with $q(z) \in A^*$, $x,y \in E$ with $q(x) = (x,y) = o$ it holds

(1.9) $$\sigma \circ \sigma_z \circ \sigma^{-1} = \sigma_{\sigma(z)}$$

(1.10) $$\sigma \circ E(x,y) \circ \sigma^{-1} = E(\sigma(x),\sigma(y))$$

In this chapter we shall throughout assume that the ring A is semi local with maximal ideals $m_1,\dots,m_s$ and Jacobson radical $r = \cap m$. The reduction M/rM of any A-module M will be denoted by $\overline{M}$ and correspondingly for any quadratic module (E,q) we shall write $(\overline{E},\overline{q})$ instead of (E(r),q(r)).

§ 2. The Eichler decomposition of the orthogonal group.

Let (M,q) be a quadratic space over A. In this section we shall throughout assume that M contains a hyperbolic plane, i.e. M contains a subspace $\langle e,f\rangle$ with $q(e) = q(f) = o$ and $(e,f) = 1$. Thus M has a decomposition

(2.1) $$M = \langle e,f\rangle \perp M_o$$

with $M_o = \langle e,f\rangle^{\perp}$. This decomposition will be fixed through this section. We now introduce some automorphisms of (M,q), which depend on (2.1). Every unit $\lambda \in A^*$ defines an element $P(\lambda) \in O(M)$ which is given by

$$P(\lambda)(e) = \lambda e,\ P(\lambda)(f) = \lambda^{-1}f,\ P(\lambda)(z) = z$$

for all $z \in M_o$. From this definition follows $P(\lambda)P(\mu) = P(\lambda\mu)$ for all $\lambda,\mu \in A^*$, that is $P(M) = \{P(\lambda) \mid \lambda \in A^*\}$ is a subgroup of O(M) and $P(M) \cong A^*$. Another important automorphism of M is defined by

$$\psi(e) = f,\ \psi(f) = e,\ \psi(z) = z \quad \text{for all } z \in M_o.$$

Then it can easily be seen that $\psi = \sigma_{e-f}$. The subgroup of $O(M)$ generated by all Siegel transvections $E(e,y), E(f,z)$ with $y,z \in M_o$ will be denoted by $\mathbb{E}(M)$. With this notations we now can state the following result (see [E], $[B]_1$).

(2.2) Lemma. Let A be a commutative Ring and (M,q) be a quadratic space over A with a decomposition (2.1). Take $\sigma \in O(M)$. If the coefficient β in $\sigma(f) = \alpha e + \beta f + t$ $(t \in M_o)$ is a unit of A, then σ can be written as follows

$$\sigma = E(e,a)\, E(f,u)\, P(\lambda)\sigma_o$$

where $y,u \in M_o$, $\lambda \in A^*$ and $\sigma_o \in O(M_o)$ are uniquely determined by σ (we identify $O(M_o)$ as a subgroup of $O(M)$ by the map $\sigma_o \to id_{<e,f>} \perp \sigma_o$)

Proof. Assume $\sigma(f) = \alpha e + \beta f + t$ $(t \in M_o)$ with $\beta \in A^*$. From $q(f) = o$ follows $q(t) = -\alpha\beta$. Hence

$$E(e,\beta^{-1}t)(f) = f + \beta^{-1}t - \beta^{-2}q(t)e$$
$$= f + \beta^{-1}t + \beta^{-1}\alpha e$$

that is $E(e,\beta^{-1}t)(f) = \beta^{-1}\sigma(f)$. Defining $\sigma' = P(\beta)\, E(e,\beta^{-1}t)^{-1}\sigma$ we obtain $\sigma'(f) = f$. Now we set $\sigma'(e) = \gamma e + \delta f + u$ with $u \in M_o$. Since $(e,f) = (\sigma'(e),\sigma'(f)) = 1$, it follows that $\gamma = 1$, and therefore $q(u) = -\delta$. Hence

$$E(f,u)(e) = e + u - q(u)f = e + u + \delta f = \sigma'(e)$$

With $\sigma_o = E(f,u)^{-1}\sigma'$ we get $\sigma_o(E) = e$, $\sigma_o(f) = f$, that is $\sigma_o \in O(M_o)$. Using the relation (1.10) we conclude $\sigma = E(e,y)\, E(f,u)\, P(\lambda)\sigma_o$ for some $y,u \in M_o$, $\lambda \in A^*$ and $\sigma_o \in O(M_o)$. Now we check the uniqueness of this decomposition. Thus let us assume

$$E(e,y)\, E(f,u)\, P(\lambda)\sigma_o = E(e,q')\, E(f,u')\, P(\lambda')\sigma_o'$$

with $y,y',u,u' \in M_o$, $\lambda,\lambda' \in A^*$, $\sigma_o,\sigma_o' \in O(M_o)$. Applying this equality to f it follows that

$$\lambda^{-1}(f + y - q(y)e) = \lambda'^{-1}(f + y' - q(y')e),$$

which implies $\lambda = \lambda'$, $y = y'$. Thus the above relation reduces to $E(f,u)\sigma_o = E(f,u')\sigma_o'$. Now we apply this relation to e and obtain $u = u'$, $\sigma_o = \sigma_o'$. This proves the lemma.

Let us now specialize this result to a semi local ring A. Then we get the following

(2.3) Theorem. Let (M,q) be a quadratic space over the semi local ring A with a decomposition (2.1). Assume rank(M) ≥ 3. Then for every $\sigma \in O(M)$ there exist an element $z \in M_o$ such that σ has a representation

$$\sigma = E(f,z)\, E(e,y)\, E(f,u)\, P(\lambda)\sigma_o$$

with $y,u \in M_o$, $\lambda \in A^*$, $\sigma_o \in O(M_o)$ uniquely determined by σ and z.

Proof. According to (2.2) we only need to show that there exists $z \in M_o$ such that the coefficient δ in $E(f,z)\,\sigma(f) = \gamma e + \delta f + u$ $(u \in M_o)$ is a unit of A. We have $\sigma(f) = \alpha e + \beta f + t$ for some $\alpha,\beta \in A$, $t \in M_o$. Hence for any $z \in M_o$ we get

$$E(f,z)\,\sigma(f) = \alpha e + [\beta - (t,z) - \alpha q(z)]\, f + t + \alpha z$$

Thus we must choose z such that

$$\beta - (t,z) - \alpha q(z) \in A^*$$

This means $\beta - (t,z) - \alpha q(z) \notin m$ for all maximal ideals m of A. For a fixed m we denote the rest classes of $\alpha,\beta,\ldots,t,z,\ldots$ in A/m and $M_o(m)$ by $\bar{\alpha},\bar{\beta},\ldots,\bar{t},\bar{z},\ldots$, respectively. Now we consider the following two cases

i) $\bar{\beta} \neq o$ in A/m. In this case we take $\bar{z} = o$.
ii) $\bar{\beta} = o$. We look for a $\bar{z} \in M_o(m)$, so that $(\bar{t},\bar{z}) + \bar{\alpha}\, q(\bar{z}) \neq o$ in A/m.

From $\bar{\sigma}(\bar{f}) = \bar{\alpha}\bar{e} + \bar{t}$ it follows that $q(\bar{t}) = o$, that is $\bar{t}$ is isotropic if $\bar{t} \neq o$. In this last case we choose $\bar{z} \in M_o(m)$ such that $q(\bar{z}) = o$

and $(\bar{t},\bar{z}) = 1$. If $\bar{t} = o$, then $\bar{\alpha} \neq o$, and we just choose $\bar{z} \in M_o(m)$ with $q(\bar{z}) \neq o$ (this is possible because $M_o(m) \neq o$ is non singular).

Thus we have constructed for every $m \in \max(A)$ an element $\bar{z}(m) \in M_o(m)$ with $\bar{\beta} - (\bar{t},\bar{z}) - \bar{\alpha}q(\bar{z}) \neq o$ in A/m. Using the chinese remainder theorem we can find an element $z \in M_o$ with $\bar{z} = \bar{z}(m)$ for all m. This element fulfills the required conditions.

(2.4) Remark. Under the assumptions of (2.3) we can write

$$O(M) = \mathbb{E}(M)\, P(M) O(M_o)$$

(2.5) Corollary. Under the hypothesis of (2.3) it follows that $\mathbb{E}(M)$ is a normal subgroup of $O(M)$.

Proof. Let σ be any element of $O(M)$. We only prove that $\sigma E(e,y)\sigma^{-1} \in \mathbb{E}(M)$ for all $y \in M_o$. According to (2.3) we have $\sigma = G P(\lambda)\sigma_o$ with $G \in \mathbb{E}(M)$, $\lambda \in A^*$, $\sigma_o \in O(M_o)$. Thus

$$\begin{aligned} \sigma E(e,y)\sigma^{-1} &= G P(\lambda)\sigma_o E(e,y)\sigma_o^{-1} P(\lambda)^{-1}G^{-1} \\ &= G P(\lambda) E(e,\sigma_o(y)) P(\lambda)^{-1}G^{-1} \\ &= G E(e,\lambda\sigma_o(y)) G^{-1} \in \mathbb{E}(M). \end{aligned}$$

The theorem (2.3) does not hold in the case $M = \langle e,f \rangle$, because the element ψ cannot be represented as in (2.3). However, if we only consider proper automorphisms of M, then the result (2.3) can be extended to $\langle e,f \rangle$ (see (3.20)).

§ 3. Proper automorphisms.

Let (M,q) be a quadratic space over the semi local ring A. In this section we shall study the subgroup of proper automorphisms of $O(M)$. Before we define this class of automorphisms of (M,q) we shall recall the corresponding definitions in the field case. Thus let us consider a field k and a quadratic space (M,q) over k. Then we distinguish two cases:

1) $Ch(k) \neq 2$. Then for every $\sigma \in O(M)$ holds $[\det(\sigma)]^2 = 1$, i.e. $\det(\sigma) = \pm 1$. We define

(3.1) $$O^+(M) = \{\sigma \in O(M) \mid \det(\sigma) = 1\}$$

In particular there is an exact sequence $1 \to O^+(M) \to O(M) \xrightarrow{\det} \{\pm 1\} \to 1$ and $O^+(M)$ is a subgroup of index 2 in $O(M)$. The elements of $O^+(M)$ are called proper automorphisms of (M,q). Setting $O^-(M) = \{\sigma \in O(M) \mid \det(\sigma) = -1\}$ we get $O(M) = O^+(M) \cup O^-(M)$.

2) $Ch(k) = 2$. In this case $\det(\sigma) = 1$ for all $\sigma \in O(M)$, that is the determinant is not an important invariant for σ. In this case one introduces another invariant, the socalled Dickson invariant, which is defined as follows (see $[D]_1$, $[D]_2$): the space (M,q) has a basis $\{e_1,f_1,\ldots,e_n,f_n\}$ with $q(e_i) = a_i$, $q(f_i) = b_i$, $(e_i,f_i) = 1$ and $M = \langle e_1,f_1\rangle \perp \ldots \perp \langle e_n,f_n\rangle$. Then every $\sigma \in O(M)$ is defined by

$$\sigma(e_i) = \sum_{j=1}^{n} (\alpha_{ij}e_j + \beta_{ij}f_j)$$

$$\sigma(f_i) = \sum_{j=1}^{n} (\gamma_{ij}e_j + \delta_{ij}f_j$$

with $\alpha_{ij},\ldots,\delta_{ij} \in k$. We define

(3.2) $$D(\sigma) = \sum_{i,j} (a_j\alpha_{ij}\gamma_{ij} + \beta_{ij}\gamma_{ij} + b_j\beta_{ij}\delta_{ij})$$

This element of k is an invariant of σ, i.e. independent of the basis $\{e_i,f_i\}$, and for any $\sigma \in O(M)$ holds $D(\sigma)^2 + D(\sigma) = 0$, that is $D(\sigma) = 0$ or 1. The map $D: O(M) \to \mathbb{Z}/2\mathbb{Z}$, $\sigma \to D(\sigma)$, is an homomorphism, and we obtain an exact sequence

$$1 \to O^+(M) \to O(M) \xrightarrow{D} \mathbb{Z}/2\mathbb{Z} \to 0$$

where $O^+(M) = \{\sigma \in O(M) \mid D(\sigma) = 0\}$. The elements of $O^+(M)$ are called proper automorphisms of (M,q). Another important property of D is the following: if (M_i,q_i) are two quadratic spaces and $\sigma_i \in O(M_i)$ $(i=1,2)$, then for $\sigma_1 \perp \sigma_2 \in O(M_1 \perp M_2)$ we have $D(\sigma_1 \perp \sigma_2) = D(\sigma_1) + D(\sigma_2)$.

(3.4) Examples. Let k be a field (of any characteristic). Let (M,q) be a quadratic space over k.

i) For any $x \in M$ with $q(x) \neq 0$ we have $\det(\sigma_x) = -1$ if $ch(k) \neq 2$ and $D(\sigma_x) = 1$ if $ch(k) = 2$.

ii) Any Siegel transvection $E(x,y)$ is proper, i.e. $E(x,y) \in O^+(M)$. In particular $\hat{\mathbb{E}}(M) \subseteq O^+(M)$.

iii) Assume $M = \langle e,f\rangle \perp M_o$, where $\langle e,f\rangle$ is the hyperbolic plane. Then ψ is not proper, whereas $P(\lambda) \in O^+(M)$ for all $\lambda \in k^*$. In particular from (ii) we deduce $\mathbb{E}(M)\,P(M)O^+(M_o) \subseteq O^+(M)$. Since $O(M) = \mathbb{E}(M)P(M)O(M_o)$ if $\dim M \geq 3$, we conclude $O^+(M) = \mathbb{E}(M)\,P(M)O^+(M_o)$. This equality is also true for $M = \langle e,f\rangle$, since it holds that $O^+(\langle e,f\rangle) = P(\langle e,f\rangle)$ (see (3.10)).

Now we want to generalize these concepts for quadratic spaces over rings. Thus let (M,q) be a quadratic space over the ring A. For any $\sigma \in O(M)$ let us denote the reduction of σ modulo $\mathfrak{m} \in \max(A)$ by $\sigma(\mathfrak{m})$.

(3.5) Definition. We set

$$O^+(M) = \{\sigma \in O(M) \mid \sigma(\mathfrak{m}) \in O^+(M(\mathfrak{m})) \text{ for all } \mathfrak{m} \in \max(A)\}$$

The elements of $O^+(M)$ are called proper automorphisms of (M,q). $O^+(M)$ is a normal subgroup of $O(M)$, nevertheless it does not have necessarily index 2 in $O(M)$, as the following example shows: let $A = k_1 \times \ldots \times k_s$ be a product of fields, so that $O(M) = O(M_1) \times \ldots \times O(M_s)$, where M_i is the reduction of M with respect to k_i. Then we have $O^+(M) = O^+(M_1) \times \ldots \times O^+(M_s)$, and therefore $[O(M) : O^+(M)] = 2^s$. More generally, if $A = A_1 \times \ldots \times A_s$ is a product of rings, then $O(M) = \prod O(M_i)$ and $O^+(M_i) = \prod O^+(M_i)$. For example if A is a semi local ring with radical $r = \cap\,\mathfrak{m}$, then

$$(3.6) \qquad O^+(\overline{M}) = \prod_{\mathfrak{m}} O^+(M_{\mathfrak{m}}).$$

(3.7) Remark. It should be noted that one can give a more intrinsic description of the proper automorphisms of (M,q) over the ring A. Let us consider $\sigma \in O(M)$. Since the map $\sigma : M \to M \subset C(M)$ has the property $(\sigma(x))^2 = q(\sigma(x)) = q(x)$, it induces an automorphism

$$C(\sigma) : C(M) \stackrel{\sim}{\to} C(M),$$

which obviously induces automorphisms $C(\sigma) : C(M)^+ \stackrel{\sim}{\to} C(M)^+$ and $C(\sigma) : D(M) \stackrel{\sim}{\to} D(M)$. If $\sigma,\tau \in O(M)$, then we have $C(\sigma\tau) = C(\sigma)C(\tau)$. On the other hand $D(M)$ is a quadratic separable algebra over A and $C(\sigma)$ is an automorphism of $D(M)$, thus it follows that $C(\sigma)$ is an involution of $D(M)$. Let now $m \in \max(A)$ be a maximal ideal of A. Localizing with respect to m we have $C(\sigma)_m : C(M)_m \stackrel{\sim}{\to} C(M)_m$, and using the identifications $C(M)_m = C(M_m)$ we obtain $C(\sigma)_m = C(\sigma_m)$. In particular $C(\sigma)$ induces the automorphism $C(\sigma_m) : D(M_m) \stackrel{\sim}{\to} D(M_m)$. Now $D(M_m) = D(M)_m$ is a quadratic separable algebra over the local ring A_m with canonical involution ρ_m = localization of the canonical involution ρ of $D(M)$, thus we conclude either $C(\sigma_m) = \rho_m$ or $C(\sigma_m) = id_{D(M_m)}$, because every automorphism of a quadratic separable algebra over a local ring has this property. Hence we can write

$$C(\sigma_m) = \rho_m^{D_m(\sigma)}$$

with $D_m(\sigma) \in \{0,1\}$. In this way we obtain a continuous map

$$D(\sigma) : \max(A) \to \mathbb{Z}/2\mathbb{Z}$$

given by $D(\sigma)(m) = D_m(\sigma)$ for all $m \in \max(A)$ (here $\max(A)$ is endowed with the usual spectral topology). The map $D(\sigma)$ is called the Dickson invariant of $\sigma \in O(M)$. The resulting map

$$D : O(M) \to C(\max(A), \mathbb{Z}/2\mathbb{Z})$$

is called the Dickson map. For any $\sigma,\tau \in O(M)$ and $m \in \max(A)$ we have

$$\rho_m^{D_m(\sigma\tau)} = C(\sigma_m\tau_m) = C(\sigma_m)C(\tau_m) = \rho_m^{D_m(\sigma)}\rho_m^{D_m(\tau)}$$

that is $D_m(\sigma\tau) = D_m(\sigma) + D_m(\tau)$. Therefore D is a group homomorphism. From the definition (3.5) we now obtain an exact sequence

$$1 \to O^+(M) \to O(M) \xrightarrow{D} C(\max(A), \mathbb{Z}/\mathbb{Z})$$

A very important result, which shall play a central role in later arguments (see for example § 5 in this chapter and its applications in chapter V) is the following one (compare $[K]_2$).

(3.8) Theorem. Let (M,q) be a quadratic space over the semi local ring A. Then the reduction map

$$\varphi_r : O^+(M) \to O^+(\overline{M})$$

is onto.

Let us first make a reduction of the problem to a special case:

(3.9) We claim that if (3.8) is true for any quadratic space (M',q') of the type (2.1), then it holds also for any quadratic space.

To prove this fact let us consider any quadratic space (M,q) over A and set $(M',q') = \langle e,f\rangle \perp M$, where $\langle e,f\rangle$ is as usual the hyperbolic plane. Our assumption says that $\varphi_r' : O^+(M') \to O^+(\overline{M}')$ is onto. On the other hand the inclusion $M \to M'$ induces an inclusion $O^+(M) \to O^+(M')$ (by $\sigma \to id_{\langle e,f\rangle} \perp \sigma$), so that the following diagram

$$\begin{array}{ccc} O^+(M) & \xrightarrow{\varphi_r} & O^+(\overline{M}) \\ \downarrow & & \downarrow \\ O^+(M') & \xrightarrow{\varphi_r'} & O^+(\overline{M}') \end{array}$$

commutes.

Let now $\overline{\sigma} \in O^+(\overline{M})$ be given. Since $\overline{\sigma} \in O^+(\overline{M}')$ and φ_r' is onto, there exists $\sigma' \in O^+(M')$ with $\varphi_r'(\sigma') = \overline{\sigma}$. Now $\overline{\sigma}(e) = e$ and $\overline{\sigma}(f) = f$ implies $\sigma'(f) = \alpha e + \beta f + t$ $(t \in M)$ with $\beta \in A^*$. Therefore we may apply the lemma (2.2) and we get a unique decomposition

$$\sigma' = E\,P(\lambda)\,\sigma$$

with $E \in \mathbb{E}(M')$, $\lambda \in A^*$, $\sigma \in O(M)$. Hence

$$\varphi_r'(\sigma') = \varphi_r(E)\varphi_r(P(\lambda))\varphi_r(\sigma) = \bar{\sigma}.$$

But this is actually the decomposition (2.2) of $\bar{\sigma}$ as an automorphism of $\bar{M}'$, thus the uniqueness of this decomposition implies $\varphi_r(\sigma) = \bar{\sigma}$. Obviously $\sigma \in O^+(M)$, because $\sigma' \in O^+(M')$, $E, P(\lambda) \in O^+(M')$. This proves that $\varphi_r : O^+(M) \to O^+(\bar{M})$ is onto. It should be noted that we never used the assumption that A is semi local, since (2.2) is valid for any ring.

Hence in the sequel we shall always assume that (M,q) hat the form $\langle e,f\rangle \perp M_o$, where $\langle e,f\rangle$ is the hyperbolic plane. Let us first treat the case $M = \langle e,f\rangle$.

(3.10) Lemma. Let A be a commutative ring. Then for $M = \langle e,f\rangle$ it holds that

$$O^+(M) = P(M) \cong A^*.$$

Proof. Take $\sigma \in O^+(M)$ and set

$$\sigma(e) = \alpha e + \beta f$$

$$\sigma(f) = \gamma e + \delta f$$

with $\alpha,\beta,\gamma,\delta \in A$. From $q(e) = q(f) = o$, $(e,f) = 1$ we deduce

$$\alpha\beta = \gamma\delta = o \quad \text{and} \quad \alpha\delta + \beta\gamma = 1$$

Consider $m \in \max(A)$. Over A/m we have $\bar{\alpha}\bar{\beta} = \bar{\gamma}\bar{\delta} = o$ and $\bar{\alpha}\bar{\delta} + \bar{\beta}\bar{\gamma} = 1$. If $\bar{\alpha} = o$, then $\bar{\beta}\bar{\gamma} = 1$ and therefore $\bar{\delta} = o$, that is $\bar{\sigma}(\bar{e}) = \bar{\beta}\bar{f}$, $\bar{\sigma}(\bar{f}) = \bar{\beta}^{-1}\bar{e}$, i.e. $\bar{\sigma} = \bar{\psi}P(\bar{\beta}) \in O^-(M(m))$, which is a contradiction. Hence $\bar{\alpha}\bar{\delta} = 1$, $\bar{\beta} = \bar{\gamma} = o$ in A/m for all $m \in \max(A)$, which implies $\alpha, \delta \in A^*$, $\beta = \gamma = o$, $\alpha\delta = 1$. In particular $\sigma = P(\alpha)$. This proves the lemma.

(3.11) Corollary. For $M = \langle e,f\rangle$ φ_r is onto.

Proof. We have $O^+(M) = P(M)$, $O^+(\bar{M}) = P(\bar{M})$. Take $\bar{\sigma} = P(\bar{\lambda})$, $\bar{\lambda} \in (A/r)^*$.

Any representative $\lambda \in A^*$ of $\bar{\lambda}$ defines $\sigma = P(\lambda)$ with $\varphi_r(\sigma) = \bar{\sigma}$.

In the sequel we shall consider a quadratic space $(M,q) = \langle e,f\rangle \perp M_o$ with $\operatorname{rank}(M_o) \geq 1$.

(3.12) Lemma. Let $y \in M_o$ be totally anisotropic, i.e. $q(y) \in A^*$ and let σ_y be the associated symmetry. Then

(3.13) $$\sigma_y = \psi P(-q(y))\, E(f,y)\, E(e,q(y)^{-1}y)\, E(f,y)$$

Proof. Let us take $z \in M_o$. Then we have $\sigma_y(z) = z - (z,y)q(y)^{-1}y$ and hence

$$E(f,y)\sigma_y(z) = z - (z,y)f - (z,y)q(y)^{-1}(y-2q(y)f)$$

$$= z + (z,y)f - (z,y)q(y)^{-1}y$$

Applying $E(e,q(y)^{-1}y)$ to this relation we get

$$E(e,q(y)^{-1}y)\, E(f,y)\sigma_y(z) = z + (z,y)f$$

that is

$$E(f,y)\, E(e,q(y)^{-1}y)\, E(f,y)\sigma_y(z) = z$$

for all $z \in M_o$. We set $F = E(f,y)\, E(e,q(y)^{-1}y)\, E(f,y)\sigma_y$. A straightforward computation shows that $F(e) = -q(y)f$ and $F(f) = -q(y)^{-1}e$, that is

$$F = P(-q(y)^{-1})\psi.$$

From this last relation and $\sigma_y^2 = id_M$ we deduce (3.13). It should be noted that the identity (3.13) is valid for any commutative ring A.

(3.14) Corollary. Let $M = \langle e,f\rangle \perp M_o$ be as in (3.12). If $S(M_o) = O(M_o)$ is valid, then

$$O(M) = \mathbb{E}(M)\, P(M) \cup \psi\, \mathbb{E}(M)\, P(M)$$

$$O^+(M) = \mathbb{E}(M)\, P(M)$$

Proof. This follows immediately from (3.13), (2.3) and (1.10).

Let us now apply this result to the field case. For a field k it is a well-known fact that O(E) = S(E) holds for every quadratic space (E,q) over k with only one exception, namely $k = \mathbb{F}_2$ and $E = \langle e_1,f_1\rangle \perp \langle e_2,f_2\rangle$, where $\langle e_i,f_i\rangle$ are two hyperbolic planes (see $[D]_1$, $[D]_2$).

(3.15) Corollary. Let k be a field and $M = \langle e,f\rangle \perp M_o$ a quadratic space over k. Then

i) if $k \neq \mathbb{F}_2$ when $M_o = \langle e_1,f_1\rangle \perp \langle e_2,f_2\rangle$ we have

$$O(M) = \mathbb{E}(M)\,P(M) \;\cup\; \psi\,\mathbb{E}(M)\,P(M)\;.$$

ii) For any field k it holds

$$O^+(M) = \mathbb{E}(M)\,P(M)\;.$$

Proof. According to (3.14) and the above remark we only need to prove (ii) in the case $k = \mathbb{F}_2$, $M_o = \langle i,j\rangle \perp \langle g,h\rangle$ where $\langle i,j\rangle$ and $\langle g,h\rangle$ are hyperbolic planes. In any case we know that $O^+(M) = \mathbb{E}(M)\,P(M)\,O^+(M_o)$. Since $k = \mathbb{F}_2$, then $P(M) = \{id\}$, that is $O^+(M) = \mathbb{E}(M)\,O^+(M_o)$. Applying (ii) to M_o we obtain $O^+(M_o) = \mathbb{E}(M_o)$, where $\mathbb{E}(M_o)$ is defined with respect (say) to $\langle i,j\rangle$. Hence $O^+(M) = \mathbb{E}(M)\,\mathbb{E}(M_o)$. Now we want to show that $\mathbb{E}(M_o) \subseteq \mathbb{E}(M)$. Since $E(i,g)$, $E(i,h)$, $E(j,g)$, $E(j,h)$ generate $\mathbb{E}(M_o)$, we only need to show that these elements belong to $\mathbb{E}(M)$. It suffices to consider $E(i,g)$. Taking

$$\tau = \sigma_{i+j}\,\sigma_{f+i+j}\,\sigma_{e+f+i}$$

we get $\tau(e) = i$, $\tau(g) = g$. Therefore

$$\tau^{-1}E(i,g)\tau = E(\tau^{-1}(i),\tau^{-1}(g)) = E(e,g) \in \mathbb{E}(M)$$

and hence $E(i,g) \in \tau\,\mathbb{E}(M)\tau^{-1} = \mathbb{E}(M)$ (see (2.5)). This proves our assertion.

(3.16) Corollary. Let $K = k_1 \times \ldots \times k_r$ be a product of fields and

$(M,q) = \langle e,f\rangle \perp M_0$ be a quadratic space over K. Then

$$O^+(M) = \mathbb{E}(M)\,P(M)$$

Proof. We have a decomposition $(M,q) = (M_1,q_1) \times \ldots \times (M_r,q_r)$ with quadratic spaces (M_i,q_i) over k_i. Hence $O^+(M) = \prod_i O^+(M_i)$, $\mathbb{E}(M) = \prod_i \mathbb{E}(M_i)$ and $P(M) = \prod_i P(M_i)$. On the other hand we have $O^+(M_i) = \mathbb{E}(M_i)\,P(M_i)$ see (3.15)), thus

$$O^+(M) = \prod_i \mathbb{E}(M_i)\,P(M_i) = \prod_i \mathbb{E}(M_i) \prod_i P(M_i)$$
$$= \mathbb{E}(M)\,P(M)\,.$$

(3.17) Proof of theorem (3.8). According to the reduction step (3.9) it suffices to consider a quadratic space $(M,q) = \langle e,f\rangle \perp M_0$ over A. Since (3.8) is true for $\langle e,f\rangle$, we may assume $\operatorname{rank}(M_0) \geq 1$. For the reduction $(\overline{M},\overline{q})$ over $A/r = \prod_m A/m$ we have shown in (3.16) that $O^+(\overline{M}) = \mathbb{E}(\overline{M})\,P(\overline{M})$, that is $O^+(\overline{M})$ is generated by the elements $E(\overline{e},\overline{y})$, $E(\overline{f},\overline{z})$ with $\overline{y},\overline{z} \in \overline{M}_0$ and $P(\overline{\lambda})$, $\overline{\lambda} \in (A/r)^*$. For any $\overline{y},\overline{z} \in \overline{M}_0$, $\overline{\lambda} \in (A/r)^*$ let us choose representatives $y,z \in M_0$, $\lambda \in A^*$. Then $E(e,y)$, $E(f,z)$, $P(\lambda)$ are liftings of the above generators, respectively. Thus every element of $O^+(\overline{M})$ can be lifted to $O^+(M)$, that is φ_r is onto. This proves the theorem.

Now we want to generalize (3.16) for any semi local ring A. To this end we shall need some results, which are proved in a more general setting in $[K]_2$. Let (E,q) be a quadratic space over A and denote the subgroup of $O^+(E)$, which is generated by evenly products of symmetries of E by $S^+(E)$. Let us quote the following useful remark from $[K]_2$:

(3.18) Remark. Let $x,y \in E$ be elements with $x - y \in rE$ and $q(x) = q(y)$. If $t \in E$ is an element, such that $q(t) \in A^*$ and $(x,t) \in A^*$, then

$$\sigma_t(x) - y \equiv -q(t)^{-1}(x,t)t \quad (\mathrm{mod}\ rE)$$

is strictly anisotropic and $\lambda(x,y,t) = \sigma_{\sigma_t(x)-y}\,\sigma_t$ maps x to y.

Moreover $\lambda(x,y,t)$ leaves the elements of $\langle x-y,t\rangle^{\perp}$ elementwise fixed. We have $\lambda(x,y,t) \in O(E,r) \cap S^{+}(E)$.

Using this sort of automorphisms of E we now can show the following (see $[K]_2$).

(3.19) Theorem. Let $F \subset E$ be a non singular subspace of E and let $\varphi : F \to E$ be an isometry, which induces modulo r the inclusion $\bar{F} \to \bar{E}$. Then φ admits an extension to an automorphism of E, which is a product of automorphisms of the form $\lambda(x,y,t)$ (as defined in (3.18)). In particular it follows that

$$O(E,r) \subseteq S^{+}(E).$$

(Under an isometry $\varphi : F \to E$ we understand a linear monomorphism φ with $q(\varphi(x)) = q(x)$ for all $x \in F$, whose image $\varphi(F)$ is a subspace of E).

Proof. We set $E = F \perp F^{\perp}$ and $n = \dim F$. The proof proceeds by induction on n. First assume $F = \langle x\rangle$. If $E = F$, then there is nothing to prove. Thus let us assume $F \neq E$, and take $t \in E$ with $q(t) \in A^{*}$, $(x,t) \in A^{*}$. Since $\varphi(x) \equiv x \pmod{rE}$, we can construct $\lambda(x,\varphi(x),t)$, which is actually an extension of φ to E. Consider now the case $n = 2$, that is $F = \langle x_1,y_1\rangle$. We can choose x_1 and y_1 such that $(x_1,y_1) = 1$, and since A is semi local we always can assume that $q(y_1) \in A^{*}$. If $F \neq E$, then there exists $t \in E$ with $q(t) \in A^{*}$, $(x_1,t) = 0$ and $(y_1,t) \in A^{*}$. Then we define $\sigma_1 = \lambda(x_1,\varphi(x_1),y_1)$ and $\sigma_2 = \lambda(\sigma_1(y_1),\varphi(y_1),\sigma_1(t))\sigma_1$. In this case σ_2 is an extension of φ to E. If $F = E$, we consider $\sigma = \lambda(x_1,\varphi(x_1),y_1)$. Then $\sigma(x_1) = \varphi(x_1)$ (see (3.18)), that is $\sigma^{-1}\varphi(x_1) = x_1$, and $\sigma^{-1}\varphi$ induces modulo r the identity. It follows that $\sigma^{-1}\varphi = id_F$. To see this we set

$$\sigma^{-1}\varphi(y_1) = \alpha x_1 + (1-2\alpha q(x_1))y_1$$

(the second coefficient is determined by $1 = (x_1,y_1) = (\sigma^{-1}\varphi(x_1),\sigma^{-1}\varphi(y_1))$. Hence

$$q(y_1) = \alpha^2 q(x_1) + \alpha(1-2\alpha q(x_1)) + (1-2\alpha q(x_1))^2 q(y_1)$$

and this implies $(1-\alpha q(x_1))(1-4q(x_1)q(y_1))\alpha = o$. Since $\alpha \in r$ and $1-4q(x_1)q(y_1) \in A^*$, we conclude that $\alpha = o$, i.e. $\sigma^{-1}\varphi(y_1) = y_1$. Assume now dim $F > 2$. Then we have $F = F_1 \perp F_2$ with dim $F_2 = 2$. Let $\varphi_1 = \varphi_{|F_1}$ be the restriction of φ to F_1. Using induction on the dimension of F we see that there exists a product σ_1 of automorphisms of the form $\lambda(x,y,t)$ such that $\sigma_{1|F_1} = \varphi_1$. Hence $\sigma_1^{-1}\varphi$ leaves F_1 elementwise fixed and we have an isometry $\sigma_1^{-1}\varphi : F_2 \to F_2 \perp F^{\perp}$. We have shown above that there exists a product σ_2 of automorphisms of the form $\lambda(x,y,t)$ such that $\sigma_{2|F_2} = \sigma_1^{-1}\varphi_{|F_2}$ (here we have extended σ_2 to F_1 through the identity, i.e. $\sigma_{2|F_1} = id_{F_1}$). Thus $\sigma_1\sigma_2$ is an extension of φ to E. This proves the first assertion of the theorem. The second one follows immediately from the first one, taking $F = E$.

Using now the inclusion $O^+(E,r) \subseteq S^+(E)$ we are able to generalize (3.16), namely we have

(3.20) Theorem. Let $(M,q) = \langle e,f\rangle \perp M_o$ be a quadratic space over the semi local ring A. Then

$$O^+(M) = \mathbb{E}(M)\,P(M)$$

Proof. Since we have treated the case $M = \langle e,f\rangle$ in (3.10), we shall assume rank$(M_o) \geq 1$. Take $\sigma \in O^+(M)$. Using the homomorphism $\varphi_r : O^+(M) \to O^+(\overline{M}) = \mathbb{E}(\overline{M})\,P(\overline{M})$ (see (3.16)) we obtain $\varphi_r(\sigma) = \overline{E}\,P(\overline{\lambda})$ for some $\overline{E} \in \mathbb{E}(\overline{M})$, $\overline{\lambda} \in (A/r)^*$. Let $E \in \mathbb{E}(M)$, $\lambda \in A^*$ be preimages of $\overline{E}$ and $\overline{\lambda}$, respectively. We set $\tau = E\,P(\lambda)$. Hence $\tau^{-1}\sigma \in O(M,r)$. According to (2.3) we can write $\tau^{-1}\sigma = G\,P(\mu)\sigma_o$ with $G \in \mathbb{E}(M)$, $\mu \in A^*$, $\sigma_o \in O(M_o)$, and since $\tau^{-1}\sigma \in O(M,r)$, it follows from the uniqueness of the above decomposition that $\sigma_o \in O(M_o,r)$. In particular $\sigma_o \in S^+(M_o)$ (see (3.19)). On the other hand the lemma (3.12) implies $S^+(M_o) \subseteq \mathbb{E}(M)\,P(M)$, so that $\sigma_o \in \mathbb{E}(M)\,P(M)$. Therefore $\tau^{-1}\sigma \in \mathbb{E}(M)\,P(M)$ and in particular $\sigma \in \mathbb{E}(M)\,P(M)$. This proves the theorem.

As an immediate consequence of (3.8) and (3.19) one deduces (see $[K]_2$).

(3.21) Theorem. Let (E,q) be a quadratic space over the semi local ring A. Then the following assertions are equivalent:

i) $O^+(E) = S^+(E)$

ii) $O^+(E(m)) = S^+(E(m))$ for all $m \in \max(A)$.

In particular, if $E(M)$ is not an orthogonal sum of two hyperbolic planes in the case $A/m = \mathbb{F}_2$ for all $m \in \max(A)$, then it holds that $O^+(E) = S^+(E)$.

We now shall apply the results of this section to give a description of the commutator $[O^+(M), O^+(M)]$ for a quadratic space $(M,q) = \langle e,f\rangle \perp M_o$ over the semi local ring A. For any $\sigma,\tau \in O^+(M)$ we write $\sigma = E\,P(\lambda)$, $\tau = F\,P(\mu)$, where $E,F \in \mathbb{E}(M)$, $\lambda,\mu \in A^*$ (see (3.20)). Then

$$\sigma\tau\sigma^{-1}\tau^{-1} = E\,P(\lambda)F\,P(\mu)\,P(\lambda^{-1})E^{-1}P(\mu^{-1})F^{-1}$$

Using (1.10) we have $P(\lambda)F = F'P(\lambda)$ and $E^{-1}P(\mu^{-1}) = P(\mu^{-1})E'$ with $E',F' \in \mathbb{E}(M)$. Hence it follows that

$$\sigma\tau\sigma^{-1}\tau^{-1} = E\,F'\,P(\lambda\mu\lambda^{-1}\mu^{-1})E'F^{-1}$$

$$= EF'E'F^{-1} \in \mathbb{E}(M).$$

Therefore we have shown that

(3.22) $$[O^+(M), O^+(M)] \subseteq \mathbb{E}(M)$$

Our next aim is to prove the equality under certain restrictions. If $M = \langle e,f\rangle$, then $[O^+(M),O^+(M)] = \mathbb{E}(M) = \{id_M\}$ (see (3.10)), so that we may assume: $\mathrm{rank}(M_o) \geq 1$. Let us now look for conditions, which ensure that the generators $E(e,y)$, $E(f,y)$ $(y \in M_o)$ of $\mathbb{E}(M)$ are commutators. If $y \in M_o$ is strictly anisotropic, then

$$E(e,y) = \sigma_{y-q(y)e}\,\sigma_y$$

and $q(y-q(y)e) = q(y)$. If either $|A/m| \geq 3$ for all $m \in \max(A)$, or

$\dim M_0 \geq 3$, one can always find $t \in M_0$ with $q(t) \in A^*$ and $(y,t) \in A^*$. Then $u = \sigma_t(y) - (y-q(y)e) = -(y,t)q(t)^{-1}t + q(y)e$ is strictly anisotropic, and we obtain (see (3.18))

$$\sigma_u\sigma_t(y) = y-q(y)e$$

Putting $\tau = \sigma_u\sigma_t \in O^+(M)$ we conclude $\sigma_{y-q(y)e} = \sigma_{\tau(y)} = \tau\sigma_y\tau^{-1}$ and therefore

$$E(e,y) = \tau\sigma_y\tau^{-1}\sigma_y = \tau\sigma_y\tau^{-1}\sigma_y^{-1}$$

is a commutator. The same holds for $E(f,y)$. Assume now $\dim M_0 \geq 5$. Then there exists $v \in M_0$, such that $(t,v) = (y,v) = o$ and $q(v) \in A^*$. In this case it follows that

$$E(e,y) = \tau\sigma_y\tau^{-1}\sigma_y^{-1} = \tau\sigma_y\sigma_v\tau^{-1}(\sigma_y\sigma_v)^{-1}$$

lies in $[O^+(M),O^+(M)]$. The same holds for $E(f,y)$. Let us now consider $E(e,y)$ with any $y \in M_0$. Under the assumption $\dim M_0 \geq 5$ we can find $y_1,y_2 \in M_0$ with $q(y_1),q(y_2) \in A^*$ and $y = y_1+y_2$ (see (3.24) below). Hence $E(e,y) = E(e,y_1)\,E(e,y_2)$, and we get again $E(e,y)$, $E(f,y)$ $\in [O^+(M),O^+(M)]$. Thus we have shown

(3.23) Theorem. Let $M = \langle e,f\rangle \perp M_0$ be a quadratic space with $\dim M_0 \geq 5$. Then

$$\mathbb{E}(M) = [O^+(M),O^+(M)]$$

To complete the proof of (3.23) we must show

(3.24) Lemma. Assume $|A/m| \geq 3$ for all $m \in \max(A)$ or $\dim M_0 \geq 3$. Then for any $y \in M_0$ there exist $y_1,y_2 \in M_0$ with $q(y_1),q(y_2) \in A^*$ and $y = y_1+y_2$.

Proof. Clearly, we may assume that A is a field. If $q(y) \neq o$ and $ch(A) \neq 2$ we simply take $y_1 = y_2 = \frac{1}{2}y$. If $q(y) \neq o$, $ch(A) = 2$ and $|A| \geq 3$ we choose $\lambda \in A$, $\lambda \neq o$, $\lambda \neq 1$ and define $y_1 = \lambda y$, $y_2 = (1-\lambda)y$.

Assume now $q(y) \neq o$, $\dim M_0 \geq 3$ and $|A| = 2$. Take a decomposition $M_o = \langle y,z\rangle \perp \langle u,v\rangle \perp \ldots$ where $q(v) \neq 0$, $(y,z) = 1$. If $q(z) \neq 0$ we take $y_1 = y+z$, $y_2 = z$. If $q(z) = 0$ we define $y_1 = y+z+v$, $y_2 = z+v$.

Let us now assume $q(y) = 0$. We distinguish two cases: if $y = 0$, we pick $z \in M_0$ with $q(z) \neq 0$ and define $y_1 = z$, $y_2 = -z$; if $y \neq 0$, then we choose $z \in M_0$ such that $q(z) = 0$, $(y,z) = 1$. In the case $|A| \geq 3$ take $\lambda \in A$, $\lambda \neq 0,1$ and define $y_1 = \lambda y + z$ and $y_2 = (1-\lambda)y - z$. If $\dim M_0 \geq 3$, then there exists $u \in \langle y,z\rangle^{\perp}$ with $q(u) \neq 0$, and we define in this case $y_1 = y+u$, $y_2 = -u$. Since the considered cases exhaust all possibilities, the proof of the lemma is complete.

Finally let us make another application of our result (3.20). Namely, we want to determine the kernel of the spinor norm of a quadratic space of the form (2.1) over a semi local ring A. First we shall recall some definitions, but we refer the reader to [Ba], $[Bo]_2$, [B] for the details.

Let (E,q) be a quadratic space over the semi local ring A. Let C(E) be the Clifford algebra of (E,q). The group

$$(3.25) \qquad \Gamma(E) = \{s \in C(E)^+ \cup C(E)^- \mid s \text{ inversible, } sEs^{-1} = E\}$$

is called the <u>Clifford group</u> of (E,q). The subgroup $\Gamma(E)^+ = \Gamma(E) \cap C(E)^+$ is called the <u>special Clifford group</u> of (E,q). Every $s \in \Gamma(E)$ defines an element $\varphi(s) \in O(E)$ by $\varphi(s)(x) = \alpha(s)xs^{-1}$ for all $x \in E$ (since $\alpha(s) = \pm s$, so $\alpha(s)xs^{-1} \in E$). Hence φ induces an homomorphism

$$\varphi : \Gamma(E) \to O(E)$$

which is onto (see [B]). Using (3.10), chap. II we see that $\mathrm{Ker}(\varphi) = A^*$, that is we have an exact sequence

$$(3.26) \qquad 1 \to A^* \to \Gamma(E) \xrightarrow{\varphi} O(E) \to 1$$

Of course φ induces a homomorphism $\varphi : \Gamma(E)^+ \to O^+(E)$ and we get an exact sequence $1 \to A^* \to \Gamma(E)^+ \to O^+(E) \to 1$. An easy computation shows that $\varphi(\alpha(s)) = \varphi(s)$ and $\varphi(\beta(s)) = \varphi(s)^{-1}$ for all $s \in \Gamma(E)$. Therefore, setting $\bar{s} = \beta \circ \alpha(s)$ for all $s \in \Gamma(E)$, we obtain

$$\varphi(s\bar{s}) = \varphi(s)\varphi(\beta \circ \alpha(s)) = \varphi(s)\varphi(\alpha(s))^{-1} = 1,$$

that is $s\bar{s} \in A^*$. Using this fact we now define two group homomorphisms $\Gamma(E) \to A^*$ and $\Gamma(E)^+ \to A^*$ by $s \to s\bar{s}$ for all $s \in \Gamma(E)$. From the above exact sequences we get two group homomorphisms

$$n : O(E) \to A^*/A^{*2}$$
$$n^+ : O^+(E) \to A^*/A^{*2}$$

which are called <u>spinor norm homomorphisms</u>. For every $\sigma \in O(E)$ is $n(\sigma) \in A^*/A^*$ its <u>spinor norm</u>. For example let us compute the spinor norm of $E(x,y) \in O^+(E)$ $(q(x) = (x,y) = o)$. It is easy to see that $E(x,y) = \varphi(1 + xy)$. Hence

$$n^+(E(x,y)) = (1 + xy)(1 + yx) = 1,$$

since $x^2 = xy + yx = o$. If $E = \langle e,f\rangle \perp E_o$ with $\langle e,f\rangle$ = hyperbolic plane, one easily sees that $n(P(\lambda)) = \lambda (\text{mod } A^{*2})$ holds for all $\lambda \in A^*$. From these examples and (3.20) we conclude

(3.26) <u>Proposition</u>. Let $E = \langle e,f\rangle \perp E_o$ be a quadratic space over A. For any $\sigma \in O^+(E)$ with $\sigma = F P(\lambda)$, $F \in \mathbb{E}(E)$, $\lambda \in A^*$, we have $n^+(\sigma) = \lambda (\text{mod } A^{*2})$.

(3.27) <u>Corollary</u>. If E contains a hyperbolic plane, then $n^+ : O(E) \to A^*/A^*$ (and in particular n) is onto.

Let us denote the kernel of n^+ by $O^+(E)'$. We assume that E contains a hyperbolic plane, that is $E = \langle e,f\rangle \perp E_o$. Then for any $\sigma = FP(\lambda) \in O^+(E)'$ it follows that $\lambda = \mu^2$ for some $\mu \in A^*$. Since the converse is also obvious, we obtain

(3.28) $$\mathbb{E}(E) \subseteq O^+(E)' = \mathbb{E}(E)\, P(E)^2,$$

where $P(E)^2 = \{P(\mu^2) \mid \mu \in A^*\}$. If $E = \langle e,f\rangle$, then $\mathbb{E}(E) = \{id_E\}$, and therefore $O^+(E) \cong A^{*2}$. Assume now $\mathrm{rank}(E_o) \geq 1$. Taking $y \in E_o$ such that $q(y) \in A^*$, one can write

$$P(\mu^2) = (\sigma_y\sigma_{e+\mu^{-1}f})(\sigma_y\sigma_{e+f})(\sigma_y\sigma_{e+\mu^{-1}f})^{-1}(\sigma_y\sigma_{e+f})^{-1}$$

because $(y,e+\mu^{-1}f) = (y,e+f) = o$, showing that $P(E)^2 \subseteq [O^+(E),O^+(E)]$. On the other hand, if $\mathrm{rank}\ (E) \geq 7$, we have shown that $\mathbb{E}(E) = [O^+(E), O^+(E)]$ (see (3.23)), thus combining this result with the above inclusion and (3.28), we obtain

(3.29) <u>Theorem</u>. Let $E = \langle e,f\rangle \perp E_o$ be a quadratic space over A with $\mathrm{rank}\ (E) \geq 7$. Then

$$O^+(E)' = \mathbb{E}(E) = [O^+(E),O^+(E)],$$

and in particular

$$O^+(E)\ /\ [O^+(E),O^+(E)] \cong A^*/A^{*2}.$$

§ 4. <u>Witt's cancellation theorem over semi local rings.</u>

Now we shall use the results of the later section to give a proof of Witt's cancellation theorem for semi local rings. There exists a large literature on this topic, but we limit the list of references to the following papers: [A], [Bak], $[D]_1$, $[K]_2$, $[Kne]_2$, [Ro], [W]. We shall give two proofs of the cancellation theorem for quadratic spaces, the first one by reduction to the field case, and the second one directly, following $[K]_2$.

In the sequel A denotes a semi local ring and (E,q) is a quadratic space over A. Let $(F,q_{|F})$ be a subspace of (E,q). An <u>isometry</u> $\varphi:F \to E$ is a monomorphism, such that $q(\varphi(x)) = q(x)$ holds for all $x \in F$, and $\varphi(F)$ is a subspace of E.

(4.1) Theorem. Every isometry $\varphi : F \to E$ can be extended to an automorphism of E, provided $q(F^{\perp}) = \{q(y) \mid y \in F^{\perp}\}$ generates the ring A.

First proof. We reduce the problem to the field case, where our proposition is well-known (see $[D]_1$). Let $\bar{\varphi} : \bar{F} \to \bar{E}$ be the reduction of φ modulo the radical r of A. The decompositions $\bar{F} = \prod_m F(m)$, $\bar{E} = \prod_m E(m)$ give a decomposition

$$\bar{\varphi} = (\varphi(m))$$

where
$$\varphi(m) : F(m) \to E(m)$$

is the reduction of φ modulo m for all $m \in \max(A)$. Using the extension theorem for the field case (see $[D]_1$), we obtain for any $m \in \max(A)$ an automorphism $\psi(m) \in O(E(m))$ with $\psi(m)_{|F(m)} = \varphi(m)$. From the hypothesis on $q(F^{\perp})$ we easily see, that we can choose $\psi(m) \in O^{+}(E(m))$, since $q(F(m)^{\perp}) \neq \{o\}$. Now define

$$\bar{\psi} = (\psi(m)) \in \prod_m O^{+}(E(m)) = O^{+}(\bar{E})$$

Using (3.8) we can find $\psi \in O^{+}(E)$ with $\varphi_r(\psi) = \bar{\psi}$. Hence $\varphi_1 = \psi^{-1} \circ \varphi : F \to E$ is an isometry, whose reduction modulo r is the natural inclusion $\bar{F} \to \bar{E}$. Since the theorem (3.19) is also true under our hypotesis on $q(F^{\perp})$ (see $[K]_2$), we can extend φ_1 to an automorphism ψ_1 of E. Therefore $\psi \circ \psi_1$ is the required extension of φ.

Second proof (after Knebusch). The proof is based on the following result (see $[B]_1$).

(4.2) Lemma. Let $M = \langle e,f \rangle \perp M_o$ be a quadratic space over A with $M_o \neq o$, $\langle e,f \rangle$ = hyperbolic plane. Then for any hyperbolic plane $\langle i,j \rangle \subset M$ (i.e. $q(i) = q(j) = o$, $(i,j) = 1$) there exists a $\sigma \in O^{+}(M)$ such that $\sigma(e) = i$, $\sigma(f) = j$.

Proof. Write $i = \alpha e + \beta f + u$ with $u \in M_o$. Using the method of the proof of (2.3) we can find $z \in M_o$ with

$$E(e,z)(i) = \alpha' e + \beta' f + u'$$

and $\alpha' \in A^*$. Replacing $\{i,j\}$ by $\{E(e,z)(i), E(e,z)(j)\}$, we may suppose $\alpha \in A^*$ in $i = \alpha e + \beta f + u$. Therefore $E(f,\alpha^{-1}u)(e) = \alpha^{-1} i$, that is

$$E(f,\alpha^{-1}u)\, P(\alpha)(e) = i$$

Now we put $G = E(f,\alpha^{-1}u)\, P(\alpha)$ and $g = G(f)$. Thus $\{G(e), G(f)\} = \{i,g\}$ is a hyperbolic pair in M. We only need to find $\tau \in O^+(M)$ such that $\tau(i) = i$, $\tau(g) = j$. Consider now the decomposition $M = \langle i,g \rangle \perp M_1$ and set $j = \gamma i + \delta g + t$ with $t \in M_1$. Since $(i,g) = (i,j) = 1$ we obtain $\delta = 1$ and hence $q(t) = -\gamma$. An easy computation shows that

$$E(i,t)(g) = g + t + \gamma i = j$$

and $E(i,t)(i) = i$. Therefore $\sigma = E(i,t)\, E(f,\alpha^{-1}u)\, P(\alpha) \in O^+(M)$ has the property $\sigma(e) = i$, $\sigma(f) = j$, proving the lemma.

Now we prove the theorem (4.1). To simplify matters we shall assume $(F, q_{|F})$ to be non singular. Thus we have $E = F \perp F^{\perp}$, and the condition on $q(F^{\perp})$ is clearly fulfilled when $F \neq E$, which is the only non trivial case. We construct the space $E \perp -F = F \perp -F \perp F^{\perp} = \mathbb{H}[F] \perp F^{\perp}$. The given isometry $\varphi : F \to E$ induces an isometry $\varphi' : F \perp -F \to E \perp -F$ by $\varphi' = \varphi \perp id_{-F}$. Now an extension $\tilde{\varphi}'$ of φ' to $E \perp -F$ leads to an extension of φ to E, because $\tilde{\varphi}' : (-F)^{\perp} \xrightarrow{\sim} (-F)^{\perp}$ since $\tilde{\varphi}'_{|-F} = id_{-F}$. But $(-F)^{\perp} = E$, thus $\tilde{\varphi}'_{|E}$ is the required extension of φ. After this remark we may suppose that F is a hyperbolic space, that is

$$F = \langle e_1, f_1 \rangle \perp \ldots \perp \langle e_n, f_n \rangle$$

with hyperbolic planes $\langle e_i, f_i \rangle$. The case $n = 1$, i.e. $F = \langle e_1, f_1 \rangle$ follows immediately from (4.2), since there exists $\sigma \in O^+(E)$ with $\sigma(e_1) = \varphi(e_1)$, $\sigma(f_1) = \varphi(f_1)$. Let us use induction on n and assume $n > 1$. Let us consider the restriction

$$\varphi' : \langle e_1, f_1 \rangle \perp \ldots \perp \langle e_{n-1}, f_{n-1} \rangle \to E$$

of φ. By induction we obtain an extension σ of φ' to E. Hence $\sigma^{-1}\varphi : F \to E$ is an isometry, which induces the identity on $\langle e_1,f_1\rangle \perp \ldots \perp \langle e_{n-1},f_{n-1}\rangle$. Therefore $\sigma^{-1}\varphi$ induces an isometry

$$\sigma^{-1}\varphi : \langle e_n,f_n\rangle \to (\langle e_1,f_1\rangle \perp \ldots \perp \langle e_{n-1},f_{n-1}\rangle)^{\perp},$$

which, as it was already proved, has an extension to an automorphism of $(\langle e_1,f_1\rangle \perp \ldots \perp \langle e_{n-1},f_{n-1}\rangle)$. We denote this extension by τ. Let us still denote by τ the extension of τ to E, which is the identity on $\langle e_1,f_1\rangle \perp \ldots \perp \langle e_{n-1},f_{n-1}\rangle$. Then $\sigma\tau$ is an extension of φ to E, proving the theorem.

(4.3) Corollary. Let E,F,G be quadratic spaces over the semi local ring A. If $E \perp F \cong E \perp G$, then $F \cong G$.

Proof. Let $\lambda : E \perp F \xrightarrow{\sim} E \perp G$ be the given isomorphism and consider the natural inclusion $i : E \to E \perp F$. We put $\alpha = \lambda \circ i : E \to E \perp G$ and extend α to an automorphism β of $E \perp G$ according to (4.1). Therefore $\gamma = \lambda^{-1} \circ \beta : E \perp G \xrightarrow{\sim} E \perp F$ has the property $\gamma(E) = E$ and consequently $\gamma : G \xrightarrow{\sim} F$.

(4.4) Remark. In general, the result (4.3) is false for bilinear spaces. A counterexample is given by the isomorphism

$$\langle 1,1,1,1\rangle \cong \langle 1,1\rangle \perp \begin{pmatrix} 0 & 1 \\ 1 & 0 \end{pmatrix}$$

over any field of characteristic 2. However there exist some special cancellation theorems for bilinear spaces over semi local rings, one of them (due to Knebusch) we quote below (see [M-H], [OM], $[K]_1$).

For any bilinear space (E,b) over the semi local ring A let V(E) be the subgroup of A, which is additively generated by the values $b(z,z)$, $z \in E$.

(4.5) Theorem. Let E,F,G be bilinear spaces over A such that $E \perp G \cong F \perp G$ and with the property $V(G) \subseteq (V(E) + 2A) \cap (V(F) + 2A)$. Then it follows that $E \cong F$, if one of the following conditions is valid:

i) A is local

ii) A is semi local, and E contains an element z with $b(z,z) = 2a$, $a \in A^*$.

This theorem follows easily from a similar result to (4.2) for bilinear spaces and the following lemma:

(4.6) Lemma. Let (E,b) be a bilinear space and M be a metabolic space over A such that $V(M) \subseteq V(E) + 2A$. Then $E \perp M \cong E \perp r \times \begin{pmatrix} 0 & 1 \\ 1 & 0 \end{pmatrix}$, where $\dim M = 2r$.

For the proof of this lemma just use example 93 : 13 of [OM] (compare also the proof of 93 : 14a in [OM]).

§ 5. Transversality theorems for quadratic spaces.

Let A be a semi local ring. According to (3.6), chap.I, any quadratic space (E,q) over A has the form $E = \langle e_1,f_1\rangle \perp \ldots \perp \langle e_n,f_n\rangle$ or $E = \langle e_1,f_1\rangle \perp \ldots \perp \langle e_n,f_n\rangle \perp \langle g\rangle$, where $(e_i,f_i) = 1$, $1 - 4q(e_i)q(f_i) \in A^*$ and $q(g) \in A^*$ for all $1 \leq i \leq n$ (of course if $\dim E = 1$, one has $E = \langle g\rangle$). We shall call such a basis (as in [B-K]) a canonical basis for E. It can easily be seen that we may suppose $q(e_i) \in A^*$ for all i. Moreover, if $|A/m| \geq 4$ for all $m \in \max(A)$, then we can also take $q(f_i) \in A^*$ for all i. We call a canonical basis of E strict canonical, if $q(e_i)$, $q(f_i) \in A^*$ for all i.

Let $\{x_1,\ldots,x_m\}$ be a basis of E. An element $x \in E$ is called transversal to this basis, if in the representation $x = \alpha x_1 + \cdots + \alpha_m x_m$ all α_i are units. Correspondingly, if $E = E_1 \perp \ldots \perp E_r$ is an orthogonal decomposition of E, we call the element $x \in E$ transversal to this decomposition, if $x = x_1 + \cdots + x_r$ with $x_i \in E_i$ and $q(x_i) \in A^*$ for $1 \leq i \leq r$.

We shall now study the following question: let $a \in A$ be an element, which is primitively represented by (E,q), that is there exists $x \in E$ with $q(x) = a$ and $\bar{x} \neq 0$ in E(m) for all $m \in \max(A)$. Now we look for

$y \in E$, which is transversal to a given basis or orthogonal decomposition of E and such that $q(y) = a$. The theorems (5.1) and (5.2) below give an answer to this question.

(5.1) Theorem. Let (E,q) be a quadratic space over A and let B be an strictly canonical basis of E. Assume $|A/m| \geq 3$ for all $m \in \max(A)$, and if $\dim E = 3$ let $|A/m| \geq 4$ for all $m \in \max(A)$. Then for any primitive element $x \in E$ there exists $\sigma \in O^+(E)$ such that $\sigma(x)$ is transversal to the basis B.

Proof. First we reduce the problem to the field case in the following way: let $\overline{E} = \prod_m E(m)$ and $O^+(\overline{E}) = \prod_m O^+(E(m))$ be the reductions of E and $O^+(E)$ modulo r, respectively. Since $x \in E$ is primitive, we have $\overline{x} \neq o$ in $E(m)$ for all $m \in \max(A)$. Now let us assume that the theorem is true for fields. Then for every $m \in \max(A)$ we can find $\sigma(m) \in O^+(E(m))$ such that $\sigma(m)(\overline{x})$ is transversal to the basis $B(m)$ of $E(m)$ ($B(m)$ is the reduction of the basis B modulo m). We define

$$\overline{\sigma} = (\sigma(m)) \in O^+(\overline{E})$$

and choose a lifting $\sigma \in O^+(E)$ of $\overline{\sigma}$ (see (3.8)). Hence $\sigma(x)$ is transversal to B over A. Therefore we may suppose that A is a field with $|A| \geq 3$ and $|A| \geq 4$ in the case $\dim E = 3$. Now let us take an element $x \neq o$ in E. If $\dim E = 1$, then there is nothing to prove. If $E = \langle e,f \rangle$ we set $x = \alpha e$ with $\alpha \neq o$. Then $\sigma = \sigma_f \sigma_e \in O^+(E)$ has the required property. If $x = \alpha f$, we just take $\sigma = \sigma_e \sigma_f$. Assume now $\dim E = 3$, $E = \langle e,f \rangle \perp \langle g \rangle$ and $A \neq \mathbb{F}_3$. We set $x = \alpha e + \beta f + g$. If $\alpha \neq o$ (or $\beta \neq o$) we may assume that $\alpha, \beta \neq o$, according to the case $\dim E = 2$, and therefore, without restriction, $x = \alpha e + \beta f$. On the other hand the relation $1 - 4q(e)q(f) \neq o$ implies either $\beta + 2\alpha q(e) \neq o$ or $\alpha + 2\beta q(f) \neq o$. Let us for example assume that $\beta + 2\alpha q(e) \neq o$. Since $A \neq \mathbb{F}_3$, we can find $\lambda \in A$ with $\lambda \neq o$, $q(e) + \lambda^2 q(g) \neq o$. Hence $z = e + \lambda g$ is anisotropic and we may define $\sigma = \sigma_g \sigma_z \in O^+(E)$. Then it follows that

$$\sigma(x) = \gamma e + \beta f + \frac{(\beta+2\alpha q(e))\lambda}{q(e)+\lambda^2 q(g)} g$$

for a suitable $\gamma \in A$. Hence the coefficients of f and g in $\sigma(x)$ are both non zero. If $\gamma = 0$, we can apply again the case dim $E = 2$, so that we can assume that $\gamma \neq 0$ holds, too. Let us now consider the case $x = \delta g$, $\delta \neq 0$. Choose $\lambda \neq 0$ with $q(e) + \lambda^2 q(g) \neq 0$ and put $z = e + \lambda g$. As in the last case, we see that the coefficients of e and g in $\sigma_z \sigma_g(x)$ does not vanish. Thus we obtain again a transversal $\sigma(x)$ to $\mathcal{B}$ with $\sigma \in O^+(E)$.

Now we consider the case dim $E = 4$, i.e. $E = \langle e_1, f_1 \rangle \perp \langle e_2, f_2 \rangle$. Using again the two dimensional case one easily shows that the only relevant case to be considered is $x = \alpha e_1 + \beta f_1$, $\alpha, \beta \neq 0$. If $A \neq \mathbb{F}_3$, then we apply the three dimensional case to finish the proof. Thus let us assume that $A = \mathbb{F}_3$. Since $1 - 4q(e_2)q(f_2) = 1 - q(e_2)q(f_2) \neq 0$, that is $q(e_2)q(f_2) \neq 1$, we must have $q(e_2) = 1$, $q(f_2) = -1$ or $q(e_2) = -1$, $q(f_2) = 1$. Correspondingly we have $q(e_1) = \pm 1$, $q(f_1) = \mp 1$. Thus we can assume that $q(e_1) = q(e_2) = 1$, $q(f_1) = q(f_2) = -1$. In particular $q(e_1 + e_2) = 2 = -1$. Therefore the coefficients of f_1 and e_2 in $\sigma_{e_1+e_2}\sigma_{e_1}(x)$ does not vanish, and hence we may apply again the two dimensional case to finish the proof.

The general situation can now easily be treated, using these special cases. This proves the theorem.

Now we prove the corresponding theorem for orthogonal decompositions of (E,q).

(5.2) <u>Theorem</u>. Let $E = E_1 \perp \ldots \perp E_s$ be an orthogonal decomposition of (E,q) with $s \geq 2$. Let x be a primitive element of E. Then there exists $\sigma \in O^+(E)$, such that $\sigma(x)$ is transversal to the above decomposition if one of the following conditions is satisfied:

i) $|A/m| \geq 4$ for all $m \in \max(A)$ and dim $E_i \geq 2$ at least for one i.

ii) $|A/m| \geq 3$ for all $m \in \max(A)$ and dim $E_i \geq 2$ at least for two different indices i.

iii) x is isotropic and s is even

iv) x is strictly anisotropic and s is odd.

<u>Proof</u>. As in the proof of (5.1) we first reduce the problem to the field case. Choosing $\sigma(m) \in O^+(E(m))$ with $\sigma(m)(\overline{x})$ transversal to the

decomposition $E(m) = E_1(m) \perp \ldots \perp E_s(m)$ for every $m \in \max(A)$, we define $\bar{\sigma} = (\sigma(m)) \in O^+(\bar{E})$. Then for any lifting $\sigma \in O^+(E)$ of $\bar{\sigma}$ we deduce, that $\sigma(x)$ is transversal to $E = E_1 \perp \ldots \perp E_s$. Hence let us assume that A is a field.

Take $x = x_1 + \ldots + x_s$ with $x_i \in E_i$, $1 \leq i \leq s$ and $x \neq o$. Clearly, it suffices to prove the assertions (i), (ii), (iii) for $s = 2$ and (iv) for $s \leq 3$. Let us first consider the case $s = 2$. We prove (i) and (ii). Since $x \neq o$ we can use similar arguments as in the proof of (5.1) to assume that $x_1, x_2 \neq o$. If $|A| \geq 4$ we distinguish two cases:
1) $q(x_1) = o$, $q(x_2) \neq o$. Since $x_1 \neq o$, there exists $y_1 \in E_1$ such that $(x_1, y_1) = 1$, $q(y_1) = o$. We choose $\lambda \in A$ with $\lambda \neq o$, $\lambda + q(x_2) \neq o$ and $\lambda + 2q(x_2) \neq o$, which is possible since we are assuming that $|A| \geq 4$. Therefore the vector $z = \lambda y_1 + x_2$ is anisotropic, and a straightforward computation shows that $\sigma_{x_2}\sigma_z(x) = x_1' + x_2'$ with $x_i' \in E_i$, $q(x_i') \neq o$ for $i = 1,2$. Thus we take $\sigma = \sigma_{x_2}\sigma_z$.
2) $q(x_1) = q(x_2) = o$. In this case we can find $y_i \in E_i$ such that $q(y_i) = o$, $(x_i, y_i) = 1$ for $i = 1,2$. Now we choose $\lambda \in A$ with $\lambda \neq o$, $\lambda + 1 \neq o$, in consequence of which we have the anisotropic vector $z = x_1 + \lambda y_1 + y_2$. Now it is easy to see that $\sigma_{x_1+y_1}\sigma_z(x) = x_1' + x_2'$ with $x_i' \in E_i$ and $q(x_2') \neq o$. Using the case (1) if necessary we finally find $\sigma \in O^+(E)$ with the required property. This finish the proof of part (i).

Now let us assume $|A| \geq 3$ and $\dim E_1$, $\dim E_2 \geq 2$. Since the case $|A| \geq 4$ was already considered, we may assume $A = \mathbb{F}_3$. In this case both, E_1 and E_2 are universal, and hence there exist $x_1' \in E_1$, $x_2' \in E_2$ with $q(x_1'), q(x_2') \neq o$ and $q(x) = q(x_1') + q(x_2') = q(x_1' + x_2')$. Using Witt's theorem we can find $\tau \in O(E)$ such that $\tau(x) = x_1' + x_2'$. One can eventually change τ to $\sigma_{x_1}\tau$ to assume that this automorphism of E is proper. Thus we have proved the assertions (i) and (ii). To prove (iii) and (iv) we may assume $A = \mathbb{F}_3$ or $A = \mathbb{F}_2$, in which case it is easy to verify both assertions. We shall omit the details.

Finally as an application of these results, we shall use it to prove a basic fact on round forms. This class of quadratic spaces will be widely studied in the next chapter. Shortly, a quadratic space (E,q) is say to be round, if any unit $\lambda \in A^*$, which is represented by (E,q) has the property $\langle\lambda\rangle \otimes E \cong E$. Then we have

(5.3) Theorem. Let (E,q) be a round quadratic space over A with $\dim E \geq 2$. Let $\rho = \langle a_1,\ldots,a_n\rangle$ be a bilinear space with $n \geq 2$. Assume one of the following conditions:

i) n is even

ii) $|A/m| \geq 3$ for all $m \in \max(A)$.

Then if $\rho \otimes q$ is isotropic, there exists a bilinear space τ over A such that

$$\rho \otimes q \cong \langle 1,-1\rangle \otimes q \perp \tau \otimes q$$

In particular, $\rho \otimes q$ contains the hyperbolic space $\dim(E) \times \mathbb{H}$. If $n = 2$, then $\rho \otimes q$ is hyperbolic.

Proof. We may assume without restriction that $a_1 = 1$, i.e. $\rho = \langle 1,a_2,\ldots,a_n\rangle$ (just scale ρ with $\langle a_1\rangle$). Hence $\rho \otimes q \cong q \perp \langle a_2\rangle \otimes q \ldots \perp \langle a_n\rangle \otimes q$. Since this space is isotropic and $\dim q \geq 2$, we use (5.2), (iii), to choose $x_1,\ldots,x_n \in q$ with $q(x_i) \in A^*$ $(1 \leq i \leq n)$ and

$$q(x_1) + a_2q(x_2) +\cdots+ a_nq(x_n) = 0.$$

Since $\langle q(x_i)\rangle \otimes q \cong q$ for $1 \leq i \leq n$ because q is round, we get

$$\langle a_2,\ldots,a_n\rangle \otimes q \cong \langle a_2q(x_2),\ldots,a_nq(x_n)\rangle \otimes q.$$

Now $a_2q(x_2) + \cdots + a_nq(x_n) = -q(x_1)$ is a unit, thus we have $\langle a_2q(x_2),\ldots,a_nq(x_n)\rangle = \langle -q(x_1)\rangle \perp \tau$ with a bilinear space τ, that is

$$\langle a_2,\ldots,a_n\rangle \otimes q \cong \langle -q(x_1)\rangle \otimes q \perp \tau \otimes q \cong \langle -1\rangle \otimes q \perp \tau \otimes q.$$

This implies $\rho \otimes q \cong q \perp \langle -1\rangle \otimes q \perp \tau \otimes q = \langle 1,-1\rangle \otimes q \perp \tau \otimes q$. Hence the theorem is proved.

In particular, if $n = 2$, we obtain the basic fact that, if the space

$<1,a> \otimes q$ is isotropic, then it is already hyperbolic (compare (2.1), chap.IV). This result for round forms q will play an important role later.

Let us tire an immediate consequence of this theorem, which was proved in [Sh-W] for fields of characteristic $\neq 2$. If p and q are two quadratic spaces over the semi local ring A, such that $p \cong \rho \otimes q$, where ρ is a bilinear space in diagonal form, we say that q divides p and write $q|p$. For every quadratic space p we denote the hyperbolic and anisotropic part by p_h and p_a, respectively. According to (4.3) they are up to isomorphism uniquely determined by p. Then we get from (5.3).

(5.4) Corollary. Let q be an anisotropic round form over the semi local ring A. Let $p = \rho \otimes q$ be as in (5.3) (i.e. $q|p$) and assume one of the conditions (i), (ii) there. Then $q|p_a$ and $q|p_h$.

CHAPTER IV

Pfister spaces over semi local rings

§ 1. Similarities

Let A be a semi local ring. Let us consider a quadratic space (E,q) and a bilinear space (M,b) over A. The set of represented values by (E,q) will be denoted by $\underline{D}(q)$ or $\underline{D}(E)$, that is

$$\underline{D}(q) = \{q(x) \mid x \in E\}.$$

Correspondingly, we set $\underline{D}(M) = \underline{D}(b) = \{b(x,x) \mid x \in M\}$. For the represented units we set $\underline{D}(q)^* = \underline{D}(q) \cap A^*$ and $\underline{D}(b)^* = \underline{D}(b) \cap A^*$. Denoting by E^* and M^* the subsets of E and M, which consist of all strictly anisotropic elements of E and M, respectively, we obtain $\underline{D}(q)^* = q(E^*)$ and $\underline{D}(b)^* = \{b(x,x) \mid x \in M^*\}$. On the other hand, a unit $\lambda \in A^*$ is called a <u>similarity norm</u> of (E,q) (resp. of (M,b)), if $(E,q) \cong \langle\lambda\rangle \otimes (E,q)$ (resp. $(M,b) \cong \langle\lambda\rangle \otimes (M,b)$). Let us denote the groups of similarity norms of (E,q) and (M,b) by N(q) and N(b), respectively. For example, $\lambda \in N(q)$ means that there exists a linear isomorphism $\sigma : E \xrightarrow{\sim} E$ such that

$$q(\sigma(x)) = \lambda q(x)$$

for all $x \in E$. In particular this formula implies $b_q(\sigma(x),\sigma(y)) = \lambda b_q(x,y)$ for all $x,y \in E$. Such a map σ is called a <u>similarity</u> of (E,q) with similarity norm $n(\sigma) = \lambda$. We denote the group of similarities of (E,q) by $\Sigma(E)$ (or $\Sigma(q)$). Hence O(E) is the subgroup of $\Sigma(E)$ of similarities of norm 1. Thus we have an exact sequence $1 \to O(E) \to \Sigma(E) \to A^*$. Similar definitions can be made for bilinear spaces.

(1.1) <u>Definition</u>. A quadratic space (E,q) over A (resp. a bilinear space (M,b)) is called <u>round</u>, if $N(q) = \underline{D}(q)^*$ (resp. $N(b) = \underline{D}(b)^*$).

Let (E,q) be a quadratic space and (E,b) be a bilinear space. We shall write $\underline{D}(E)$ for $\underline{D}(q)$ or $\underline{D}(b)$ and N(E) for N(q) or N(b), respectively. For every $x \in E$ let n(x) denote q(x) in the quadratic case and b(x,x) in the bilinear case. If $1 \in \underline{D}(E)$, then $N(E) \subseteq \underline{D}(E)^*$, because from $\sigma: \langle\lambda\rangle \otimes E \tilde{\to} E$ and $n(x) = 1$ for some $x \in E$ it follows that $n(\sigma(x)) = \lambda$. Now we claim: E is round if and only if $D(E)^* \subseteq N(E)$. Let us assume $\underline{D}(E)^* \subseteq N(E)$. According to the above remark, it suffices to prove $1 \in \underline{D}(E)^*$. Now for any $x \in E$ with $n(x) \in A^*$ there exists a similarity σ of E with norm $n(x)$, that is $n(\sigma(z)) = n(x)n(z)$ for all $z \in E$. In particular for $z = x$ we get $n(n(x)^{-1}\sigma(x)) = 1$, i.e. $1 \in \underline{D}(E)^*$.

(1.2) Example. Let us consider the quadratic space $[1,a]$ with $1-4a \in A^*$. We claim that $[1,a]$ is round. Correspondingly, for any unit $b \in A$ the bilinear space $\langle 1,b\rangle$ is round. To see this fact, let us recall that $[1,a]$ is the norm form of $B = A(\wp^{-1}(-a))$, that is $[1,a] = (B,n)$ with $n(\alpha) = \alpha\bar{\alpha}$ = norm of α. Thus we have $\underline{D}([1,a])^* = \{\alpha\bar{\alpha} \mid \alpha \in B^*\}$. On the other hand, every $\alpha \in B^*$ defines a similarity $\hat{\alpha}: B \to B$ by $\hat{\alpha}(\beta) = \alpha\beta$ for all $\beta \in B$, with norm $n(\alpha)$ (because $n(\alpha\beta) = n(\alpha)n(\beta)$), that is we have $\underline{D}([1,a])^* \subset N([1,a])$, proving that $[1,a]$ is round. One can prove this fact directly by giving explicitely the similarity with norm $q(x) \in A^*$ for any $x \in [1,a]$. Namely, if $[1,a] = Ae \oplus Af$ with $q(e) = 1$, $q(f) = a$, $(e,f) = 1$ and $x = \gamma e + \delta f$ ($q(x) \in A^*$), we define $\sigma: [1,a] \to [1,a]$ by the matrix

$$\begin{pmatrix} \gamma & \delta \\ -a\delta & \gamma+\delta \end{pmatrix}$$

Let us now consider $\langle 1,b\rangle = Ag \oplus Ah$ with $(g,g) = 1$, $(g,h) = 0$, $(h,h) = b \in A^*$. Take $x = \gamma g + \delta h$ with $(x,x) = \gamma^2 + \delta^2 b \in A^*$. Then we define a similarity $\sigma: \langle 1,b\rangle \to \langle 1,b\rangle$, whose norm is (x,x), by

$$\sigma = \begin{pmatrix} \gamma & \delta \\ -b\delta & \gamma \end{pmatrix}$$

This proves that $\langle 1,b\rangle$ is round.

(1.3) Remark. Now we introduce another point of view, which is inspired on the work of Shapiro (see [Sh]). Let (E,q) be a quadratic space over the semi local ring A. We call an endomorphism $\sigma: E \to E$ a similarity endomorphism with similarity norm λ, if $q(\sigma(x)) = \lambda q(x)$ holds for all $x \in E$. One easily sees that $\lambda \in A^*$ is equivalent with $\sigma \in \Sigma(E)$. We write $n(\sigma)$ for the similarity norm of σ, and we denote the set of all similarity endomorphisms of (E,q) by $Sim(E)$. The map $n: Sim(E) \to A$ has the property $n(\sigma\tau) = n(\sigma)n(\tau)$, $n(a\sigma) =$

$a^2 n(\sigma)$ for all $\sigma, \tau \in Sim(E)$, $a \in A$. For any $\sigma \in End(E)$ we denote the adjoint endomorphism of σ with respect to b_q by $\tilde{\sigma}$, that is

$$b_q(\sigma(x), y) = b_q(x, \tilde{\sigma}(y))$$

for all $x, y \in E$. Then it is very easy to see that $\sigma \in Sim(E)$ if and only if $\tilde{\sigma}\,\sigma = c.1_E$ for some $c \in A$. In this case it holds $n(\sigma) = c$. Using this property one easily proves that the following statements are equivalent for any pair of elements $\sigma, \tau \in Sim(E)$:

i) $\sigma + \tau \in Sim(E)$

ii) $a\sigma + b\tau \in Sim(E)$ for all $a, b \in A$

iii) $\tilde{\sigma}\tau + \tilde{\tau}\sigma = c.1_E$ for some $c \in A$.

Now this fact implies that the restriction of n to any submodule F of $Sim(E)$ defines a quadratic form on F. In this case for $\sigma, \tau \in F$ we have $\tilde{\sigma}\tau + \tilde{\tau}\sigma = b_n(\sigma, \tau) 1_E$. This point of view conduces to a geometrical interpretation of the composition of two quadratic spaces. Let us first define this notion:

(1.4) Definition. Let (E,q), (F,p) be two quadratic spaces over A. If there exists a map $c: F \times E \longrightarrow E$ such that

$$q(c(u,x)) = p(u)\, q(x)$$

for all $u \in F$, $x \in E$, we say that (F,p) and (E,q) admit a composition. We say that the composition is semi-linear, if c is linear on the second variable. We say that the composition is linear, if c is bilinear. In the sequel we write $u \cdot x$ instead of $c(u,x)$ for $u \in F$, $x \in E$.

Now let us assume that F and E have a semi linear composition. Then for any $u \in F$ we define $c(u): E \longrightarrow E$ by $c(u)(x) = u \cdot x$ for all $x \in E$. From the definition above we see that $c(u) \in Sim(E)$ for all $u \in F$. Thus we get a map $c: F \longrightarrow Sim(E)$. Of course the similarity norm of $c(u)$ is $p(u)$, thus we conclude that $c: F^* \longrightarrow \Sigma(E)$. The map c is linear if and only if the composition c is linear.

Now we want to show that $c : F \longrightarrow Sim(E)$ is one to one. To this end we need one of the following two formulas:

(1.4) $b_q(u \cdot z, u \cdot y) = p(u)\, b_q(z, y)$

(1.5) $b_q(u\cdot z,\ v\cdot z) = b_p(u,v)\ q(z)$

for all $u,v \in F$, $z,y \in E$. To prove (1.5) we just note that $b_q(u\cdot z,\ v\cdot z) = q((u+v)\cdot z) - q(u\cdot z) - q(v\cdot z) = (p(u+v) - p(z) - p(v))\ q(z) = b_p(u,v)\ q(z)$. Now assume that $c(u) = 0$ for some $u \in F$, that is $u\cdot z = 0$ for all $z \in E$. Choosing $z \in E$ with $q(z) \in A^*$ we conclude from (1.5) that $b_p(u,v) = 0$ for all $v \in F$, and hence $u = 0$. If the composition c is linear, we have proved $(F,p) \cong (c(F),\ n_{|c(F)}) \subseteq \mathrm{Sim}(E)$. Using this last isomorphism we can identify (F,p) with a quadratic subspace of $\mathrm{Sim}(E)$. Thus to say that there exists a linear composition $c : F \times E \longrightarrow E$ is equivalent with the choise of a quadratic space $(F,p) \subseteq \mathrm{Sim}(E)$. In the linear case we also have the fact, that (F,p) is similar to a subspace of (E,q). To prove this, let us take $z \in E$ with $q(z) \in A^*$ and define $\sigma : F \longrightarrow E$ by $\sigma(u) = u\cdot z$ for all $u \in F$. As above, we immediately see that σ is injectiv. Since $q(\sigma(u)) = q(uz) = p(u)\ q(z)$, it follows that σ is a similarity with norm $q(z)$, that is $\langle q(z)\rangle \otimes (F,p) \cong (\sigma(F),q)$. In particular we obtain $\dim F \leq \dim E$ (see [Sh], (1.9)). The Hurwitz problem now reads: determine the quadratic spaces (E,q) of dimension n, which admit a linear composition with a quadratic space (F,p) with $\dim F = \dim E$. Now if $c : F \times E \longrightarrow E$ is a linear composition, we can normalize c, so that $1_E \in c(F)$. To see this we choose $v \in F$ with $p(v) \in A^*$ and define $\bar{c} : F \times E \longrightarrow E$ by $\bar{c}(u,y) = c(v)^{-1}\ (u\cdot y)$ for all $u \in F$, $y \in E$. Then $\bar{c}(v) = 1_E$. In the sequel we assume that c is normalized in this sense, and hence $1_E \in F \subset \mathrm{Sim}(E)$ (here we have identified F with c(F)). Thus we may write $F = [1,b] \perp F'$ where $[1,b] = \langle 1_E, f\rangle$ and $(1_E,f) = 1$, $n(f) = b$. For any $g \in F'$ the relation $(1_E,\ g) = 0$ implies $\tilde{g} + g = 0$ and hence $g^2 = -n(g)$. The relation $(1_E,f) = 1$ means $\tilde{f} + f = 1_E$, i.e. $\tilde{f} = 1_E - f$ and since $\tilde{f}f = b$, we obtain $f^2 = f - b1_E$. Using these relations we obtain two representations

$$\rho_1 : A(\wp^{-1}(-b)) \longrightarrow \mathrm{End}\ (E)$$

$$\rho_2 : C(-F') \longrightarrow \mathrm{End}\ (E)$$

If $A(\wp^{-1}(-b)) = A \oplus Az$ with $z^2 = z - b$, then ρ_1 is defined by $\rho_1(z) = f$. The representation ρ_2 can be defined using the universal property of Clifford algebras. On the other hand for all $u \in [1,b]$ $g \in F'$ we have $\tilde{u}g + \tilde{g}u = 0$, and therefore $\tilde{u}g = gu$. Since the involution ~ correspond to the canonical involution of $A(\wp^{-1}(-b))$

(because $\tilde{f} = 1_E - f$) we obtain a representation

$$(1.6) \qquad \rho : A(\wp^{-1}(-b)) \hat{\otimes}_A C(-F') \longrightarrow \text{End}(E)$$

This representation was first constructed by Hurwitz in [H] (see (9), (10) there) and recently Shapiro has exploited it thoroughly for fields of characteristic $\neq 2$ in his work [Sh] (compare also [L]). Let us set $m = \text{rank}(F)$ and $n = \text{rank}(E)$. Now using (3.5), chap. II, (1.6) above and theorem (4.3), chap. II of [I - de M] we conclude that

$$(1.7) \qquad 2^{m-2} \mid n^2$$

This correspond to the relation (13) of [H] and to (3.3) in [Sh]. Hence if $n = 2^t n_o$ with $2 \nmid n_o$, then $m - 2 \leq 2t$. Let us set $\rho(n)$ for the maximal rank of quadratic subspaces (F,p) of Sim (E), when E ranges over all quadratic spaces of rank n over A. This function $\rho(n)$ is called the <u>Hurwitz function</u> for semilocal rings. Thus, the above remarks show:

(1.8) <u>Proposition</u>. For any natural number $n \geq 1$ with $n = 2^t n_o$, $2 \nmid n_o$ it follows that $\rho(n) \leq 2t + 2$.

In particular if we look for n such that $\rho(n) = n$, we must have $2^t n_o \leq 2t + 2$. One easily sees that this can only happen if $n = 1, 2, 4$ or 8. Thus the Hurwitz problem can be solved only for the dimensions 1, 2, 4 and 8. In the next section we shall construct quadratic spaces of these dimensions, which solve the Hurwitz problem.

Finally let us remark, how we can redefine the round spaces using the above notions. First we change a little the definition (1.4). We say that the quadratic spaces (F,p), (E,q) have a <u>restricted composition</u>, if there exists amap $F^* \times E \longrightarrow E$ with $q(u \cdot x) = p(u)q(x)$ for all $u \in F^*$, $x \in E$, which is linear in the second variable. Hence every semi linear composition induces a restricted composition. Using this notion we obtain.

(1.9) <u>Proposition</u>. A quadratic space (E,q) is round if and only if (E,q) has a restricted composition with himself.

<u>Proof</u>. If (E,q) is round, then $\underline{D}(E)^* = N(E)$. Hence for any $u \in E^*$ there exists $\sigma \in \Sigma(E)$ with norm $n(\sigma) = q(u)$. We define $u \cdot x = \sigma(x)$

for all $x \in E$. This defines a restricted composition $E^* \times E \longrightarrow E$. Conversely, if we have a restricted composition $E^* \times E \longrightarrow E$, then $\sigma(x) = u.x$ defines for all $u \in E$ a similarity of E with norm $q(u)$, that is $\underline{D}(E)^* \subset N(E)$. This proves that E is round.

§ 2. Pfister spaces.

Let us begin with a result, which will enable us to construct a lot of different round spaces starting from a given one.

(2.1) Theorem. Let A be a semi local ring with $|A/m| \geq 3$ for all $m \in \max(A)$. Let (E,q) be a round quadratic space over A and $a \in A^*$. Then

$$F = \langle 1,a\rangle \otimes E$$

is round.

Proof. We set $F = E \perp \langle a\rangle\otimes E$ and $\langle a\rangle = At$ with $(t,t) = a$. If dim E $= 1$, then $E = \langle 1\rangle$ and $F = \langle 1,a\rangle$ is round in virtue of (1.2) ($\langle 1\rangle$ is the only possible one dimensional round space). We now assume that dim $E \geq 2$. Every element of F has the form $x + t\otimes y$ with $x,y \in E$ and $q(x + t\otimes y) = q(x) + aq(y)$. Hence assuming $q(x) + aq(y) \in A^*$, we have to show that

$$\langle q(x) + aq(y)\rangle \otimes F \cong F .$$

But we know that $x + t\otimes y$ is primitive, thus we can find $x',y' \in E$ with $q(x' + t\otimes y') = q(x + t\otimes y)$ and $q(x'), q(y') \in A^*$ (use (5.2)(ii), chap.III). Therefore we can assume that $q(x), q(y) \in A^*$. Since E is round, we have $\langle q(x)\rangle \otimes E \cong \langle q(y)\rangle \otimes E \cong E$, and hence $\langle 1,a\rangle \otimes E \cong \langle 1,aq(x)q(y)\rangle \otimes E$. This implies

$$\langle q(x) + aq(y)\rangle \otimes F \cong \langle q(x) + aq(y)\rangle \otimes \langle 1,aq(x)q(y)\rangle \otimes E .$$

On the other hand for any pair of units $\lambda, \mu \in A^*$ with $\lambda + \mu \in A^*$ we have

$$\text{(2.2)} \qquad \langle \lambda,\mu\rangle \cong \langle \lambda + \mu\rangle \otimes \langle 1,\lambda\mu\rangle ,$$

thus it follows for $\lambda = q(x)$, $\mu = aq(y)$ that

$$\begin{aligned} \langle q(x) + aq(y)\rangle \otimes F &\cong \langle q(x), aq(y)\rangle \otimes E \\ &\cong \langle 1,a\rangle \otimes E \\ &\cong F \end{aligned}$$

proving the theorem.

(2.3) Remark. The following construction will enable us to reduce some problems to semi local rings, whose residue class fields are not too small. Let A be a semi local ring and define

$$B = A[X]/(X^3 + 6X^2 - X + 1).$$

Of course B is a semi local ring. Let x denote the class of X in B. Then $B = A \oplus Ax \oplus Ax^2$ with $x^3 = -6x^2 + x - 1$. The residue class fields B/M, $M \in \max(B)$ have at least 7 elements. To see this let us set $m = A \cap M$ for some $M \in \max(B)$. If $|A/m| \geq 7$, then there is nothing to prove. If $|A/m| < 7$, then the polynomial X^3+6X^2-X+1 is irreducible over A/m, that is B/M is an extension of degree 3 over A/m, which implies $|B/M| > 7$. Defining the trace map $s\colon B \longrightarrow A$ by $s(1) = 1$, $s(x) = s(x^2) = 0$, we easily see that (B,s) is a Frobenius extension of degree 3 over A.

(2.4) Theorem. For $a_1, \ldots, a_n \in A^*$ and $b \in A$ with $1-4b \in A^*$, the quadratic space

$$q = \langle 1,a_1\rangle \otimes \ldots \otimes \langle 1,a_n\rangle \otimes [1,b]$$

is round over any semi local ring A.

Proof. Since $[1,b]$ is round (see (1.2)), the assertion of the theorem follows by induction in the case that $|A/m| \geq 3$ for all $m \in \max(A)$. We now consider any semi local ring. Using the cubic extension B of remark (2.3) we conclude that $q \otimes B$ is round. Consider $\lambda \in \underline{D}(q)^*$. In particular $\lambda \in \underline{D}(q \otimes B)^*$ and hence

$$q \otimes B \cong \langle\lambda\rangle \otimes (q \otimes B)$$

Taking the trace s_* of this space we obtain

$$s_*(q \otimes B) \cong \langle\lambda\rangle \otimes s_*(q \otimes B)$$

and since $s_*(q \otimes B) = s_*(\langle 1\rangle) \otimes q$, we get

$$s_*(\langle 1\rangle) \otimes q \cong \langle\lambda\rangle \otimes s_*(\langle 1\rangle) \otimes q.$$

On the other hand we have

(2.5) $$s_*(\langle 1\rangle) \cong \langle 1\rangle \perp \begin{pmatrix} 0 & 1 \\ 1 & 6 \end{pmatrix}$$

that is

$$q \perp \mathbb{H}[q] \cong \langle\lambda\rangle \otimes q \perp \mathbb{H}[q]$$

Now the cancellation theorem (4.3), chap. III, implies

$$q \cong \langle\lambda\rangle \otimes q$$

i.e. q is round. This proves the theorem.

(2.6) Remark. The theorem (2.1) is also true for bilinear spaces, that is, for any round bilinear space (M,b) over the semi locar ring A with $|A/m| \geq 3$ for all $m \in \max(A)$ and any unit $a \in A^*$ the space $\langle 1,a\rangle \otimes M$ is round, too (see $[K]_3$). We shall use this fact only for bilinear spaces of the form $\langle 1,a_1\rangle \otimes \ldots \otimes \langle 1,a_n\rangle$, in which case we prove below a much more precise result (see (2.8) below). It should be noted that the method of proof of (2.4) does not apply in the general case of theorem (2.1), since it is not clear, if for a round space E over A the extended space $E \otimes B$ is round. It would be desirable to find a proof of (2.1) for any semi local ring.

(2.7) Definition. Take $a_1, \ldots, a_n \in A^*$, $b \in A$ with $1-4b \in A^*$. The quadratic space $q = \langle 1,a_1\rangle \otimes \ldots \otimes \langle 1,a_n\rangle \otimes [1,b]$ is called a (n+1)-fold quadratic Pfister space. This space will be denoted by

$$\langle\langle a_1, \ldots, a_n, b]]$$

The bilinear space $\varphi = \langle 1,a_1\rangle \otimes \ldots \otimes \langle 1,a_n\rangle$ is called a n-fold bilinear Pfister space. It will be denoted by $\langle\langle a_1, \ldots a_n\rangle\rangle$ (this last notation has been introduced by Elman and Lam in $[E\text{-}L]_3$).

It can easily be seen that the Pfister spaces $<1>$, $[1,b]$, $<<a_1,b]]$, $<<a_1,a_2,b]]$ solve the Hurwitz problem for the dimensions $n = 1, 2, 4$ and 8, respectively.

The rest of this section is devoted to the study of bilinear Pfister spaces. For the 1-fold Pfister space $<1,a>$ with $a \in A^*$ we choose a basis $\{z_1,z_2\}$ with $(z_1,z_1) = 1$, $(z_1,z_2) = 0$, $(z_2,z_2) = a$. Take now a n-fold Pfister space $\varphi = <1,a_1> \otimes \ldots \otimes <1,a_n>$ with $a_i \in A^*$. Tensoring up all such basis for each $<1,a_i>$ $(1 \leq i \leq n)$ we obtain a basis $\{z_1,\ldots z_{2^n}\}$ of φ with $\varphi(z_1) = 1,\ldots,$ $\varphi(z_{2^n}) = a_1\ldots a_n$, $\varphi(z_i,z_j) = 0$ for $i \neq j$ (here we write $\varphi(x)$ instead of $\varphi(x,x)$ for all $x \in \varphi$). Such a basis of the bilinear Pfister space φ will be called a typical basis of φ. The element $z_1 \in \varphi$ with $\varphi(z_1) = 1$ induces a decomposition $\varphi = <z_1> \perp <z_2,\ldots,z_{2^n}> = <1> \perp \varphi'$. We shall later show that φ' is uniquely determined (up to isomorphism) by the relation $\varphi \cong <1> \perp \varphi'$. With these notations we have

(2.8) <u>Theorem</u>. Let A be a semi local ring with $|A/m| \geq 3$ for all $m \in \max(A)$. Let $\varphi = <1,a_1> \otimes \ldots \otimes <1,a_n> = <z_1> \perp \varphi'$ be a bilinear n-fold Pfister space over A. Then for any $z \in \varphi$ with $\varphi(z) \in A^*$ there exists a similarity $\sigma : \varphi \longrightarrow \varphi$ with norm $\varphi(z)$ such that

$$\sigma(z) = \varphi(z)z_1$$

In particular φ is round.

(2.9) <u>Remark</u>. The condition $\sigma(z) = \varphi(z)z_1$ implies for any $\omega \in \varphi$ the relation

$$(2.10) \qquad \sigma(\omega) = (\omega,z)z_1 + \omega'$$

with $\omega' \in \varphi'$. Namely if $\sigma(\omega) = \alpha z_1 + \omega'$ for some $\alpha \in A$, $\omega' \in \varphi'$, we obtain from $\sigma(z) = \varphi(z)z_1$

$$\varphi(z)(\omega,z) = (\sigma(\omega),\sigma(z)) = \alpha\varphi(z)$$

that is $(\omega,z) = \alpha$. Now applying φ to (2.10) it follows that

$$(2.11) \qquad \varphi(z)\varphi(\omega) = (z,\omega)^2 + \varphi(\omega')$$

Here the components of $\omega' \in \varphi' = <z_2,\ldots z_{2^n}>$, with respect to the basis $z_2, \ldots z_{2n}$, are linear functions in the components of ω and rational functions in the components of z (compare [Pf]$_2$). This generalizes a well-known formula of Pfister (loc.cit.).

For the proof of (2.8) we need two lemmas, which hold over any commutative ring A.

(2.12) Lemma. Let $\gamma, \delta \in A^*$, $\xi, \eta \in A$ be elements such that $\xi^2\gamma + \eta^2\delta \in A^*$. Consider the bilinear spaces $Ax_1 \perp Ax_2 = \langle\gamma,\delta\rangle$ and $Ay_1 \perp Ay_2 = \langle 1,\gamma\delta\rangle$. Then the map $\Theta : Ay_1 \perp Ay_2 \longrightarrow Ax_1 \perp Ax_2$, given by $\Theta(y_1) = \xi x_1 + \eta x_2$, $\Theta(y_2) = \eta\delta x_1 - \xi\gamma x_2$, defines a linear isomorphism such that

$$(\Theta(u),\Theta(\nu)) \quad = \quad (\xi^2\gamma + \eta^2\delta)(u,\nu)$$

for all $u,\nu \in Ay_1 \perp Ay_2$.

Proof. By direct computation.

(2.13) Lemma. Let F be a bilinear or quadratic space over A. Consider $E = \langle 1,a\rangle \otimes F = F \perp \langle t\rangle \otimes F$ (where $At = \langle a\rangle$, $(t,t) = a \in A^*$) for some $a \in A^*$. Let σ, τ, τ' be similarities of F with norms λ, μ, μ^{-1}, respectively. Then for any $\xi, \eta \in A$ with $\xi^2\lambda + \alpha\eta^2\mu \in A^*$ the map

$$\Phi : E \longrightarrow E$$

given by the matrix (with respect to $F \perp \langle t\rangle \otimes F$)

$$\Phi = \begin{pmatrix} \xi\sigma & \eta\tau \\ \eta\mu\alpha\tau' & -\xi\lambda\tau\sigma^{-1}\tau' \end{pmatrix} \tag{2.14}$$

defines a similarity of E with norm $\xi^2\lambda + \alpha\mu\eta^2$.

Proof. Take $\langle\lambda,a\mu\rangle = Ax_1 \perp Ax_2$ and $\langle 1,a\lambda\mu\rangle = Ay_1 \perp Ay_2$. The linear isomorphism

$$\alpha : F \perp \langle t\rangle \otimes F \longrightarrow \langle y_1,y_2\rangle \otimes F$$

given by $\alpha(\nu) = y_1 \otimes \nu$, $\alpha(t \otimes \omega) = y_2 \otimes \sigma^{-1}\tau'$ for all $\nu,\omega \in F$ is an isomorphism for the corresponding forms. The linear isomorphism

$$\beta : \langle y_1,y_2\rangle \otimes F \longrightarrow \langle x_1,x_2\rangle \otimes F$$

given by $\beta = \Theta \otimes \mathrm{id}_F$ is a similarity with norm $\xi^2\lambda + a\mu\eta^2$ (see (2.12)). Finally, the linear isomorphism

$$\gamma: \langle x_1,x_2\rangle \otimes F \longrightarrow F \perp \langle t\rangle \otimes F$$

given by $\gamma(x_1 \otimes v) = \sigma(v)$, $\gamma(x_2 \otimes w) = t \otimes \tau(w)$ for $v,w \in F$ is an isomorphism for the corresponding forms.
Hence $\Phi = \gamma \circ \beta \circ \alpha : E \longrightarrow E$ is a similarity with norm $\xi^2\lambda + a\mu\eta^2$, which, as one easily sees, is given by the matrix (2.14). This proves the lemma.

(2.15) Remark. It should be noted that if we drop in (2.3) the assumption $\xi^2\lambda + a\mu\eta^2 \in A^*$, we obtain a similarity endomorphism of E i.e. $\Phi \in \mathrm{Sim}(E)$. Denoting by $n(u)$ the value of the quadratic or bilinear form of F on $u \in F$, we get for all $v,w \in F$

$$(\xi^2\lambda + a\mu\eta^2)(n(v) + an(w)) = n(\xi\sigma(v) + \mu\eta a\tau'(w)) + an(\eta\tau(v) - \xi\lambda\tau\sigma^{-1}\tau'(w))$$

Now we return to the proof of (2.8).

Proof of (2.8). The proof proceeds by induction on n. First assume $n = 1$, i.e. $\varphi = \langle 1,a\rangle = \langle z_1,z_2\rangle$ and consider $z = \xi z_1 + \eta z_2 \in \varphi$ with $\varphi(z) = \xi^2 + a\eta^2 \in A^*$. We define $\sigma: \varphi \longrightarrow \varphi$ by $\sigma(z_1) = \xi z_1 + \eta z_2$, $\sigma(z_2) = a\eta z_1 - \xi z_2$ (see (2.12)), which is a similarity with norm $\varphi(z)$. From the definition of σ it follows that $\sigma(z) = \varphi(z)z_1$.
We now assume $n > 1$. We set $\psi = \langle 1,a_2\rangle \otimes \ldots \otimes \langle 1,a_n\rangle$ and hence $\varphi = \langle 1,a\rangle \otimes \psi = \psi \perp \langle t\rangle \otimes \psi$ with $At = \langle a\rangle$, $(t,t) = a$ (here we set $a = a_1$). Let $\{z_1,\ldots,z_{2^{n-1}}\}$ be a typical basis for ψ. Then $\{z_1,\ldots,z_{2^{n-1}},t\otimes z_1,\ldots,t\otimes z_{2^{n-1}}\}$ is a typical basis of φ. Any $z \in \varphi$ has the form $z = x + t\otimes y$ with $x,y \in \psi$ and $\varphi(z) = \psi(x) + a\psi(y)$. Let us assume $\varphi(z) \in A^*$ and put $b = \psi(x)$, $c = \psi(y)$, so that $\varphi(z) = b + ac$. Now we distinguish two cases:

i) $b, c \in A^*$. Using the induction hypothesis, we can find similarities $\sigma: \psi \longrightarrow \psi$, $\tau: \psi \longrightarrow \psi$ with norms $b = \psi(x)$, $c = \psi(y)$, such that $\sigma(x) = bz_1$ and $\tau(y) = cz_1$, respectively. Similarly, for $c^{-1} = \psi(c^{-1}y)$ there exists a similarity $\tau': \psi \longrightarrow \psi$ with norm c^{-1}, such that $\tau'(c^{-1}y) = c^{-1}z_1$, that is $\tau'(y) = z_1$. We now apply the lemma

(2.12) to the similarities σ, τ, τ' with $\xi = \eta = 1$, and we obtain a similarity $\Phi: \varphi \longrightarrow \varphi$ with norm $b + ac$, which is given by

$$\Phi(v + t\otimes w) = \sigma(v) + ac\,\tau'(w) + t\otimes [\tau(v) - b\tau\,\sigma^{-1}\tau'(w)]$$

for all $v, w \in \psi$. Using $\sigma(x) = bz_1$, $\tau(y) = cz_1$ and $\tau'(y) = z_1$, we conclude for $z = x + t\otimes y$ that

$$\Phi(z) = (b + ac)z_1 = \varphi(z)z_1 ,$$

proving our assertion in this case.

ii) Assume that either b or c is not a unit. Let us consider elements $g, h \in A$, $u, v \in \psi$ such that $g^2\psi(u) + h^2a\psi(v) \in A^*$ and $\psi(u)$, $\psi(v) \in A^*$. The induction hypothesis implies that there exist similarities σ, τ, τ' of ψ with norms $\psi(u)$, $\psi(v)$ and $\psi(v)^{-1}$, respectively. The formula in remark (2.15) reads now

$$(g^2\psi(u) + ah^2\psi(v))(\psi(x) + a\psi(y)) = \psi(x') + a\psi(y')$$

where $x' = g\sigma(x) + h\psi(v)a\tau'(y)$ and $y' = h\tau(x) - g\psi(u)\tau\sigma^{-1}\tau'(y)$. We define $z' = x' + t\otimes y' \in \varphi$. Now we need to find g, h, u, v as above with the additional property

$$\psi(x'),\ \psi(y') \in A^* .$$

For example, if $|A/m| \neq 2,3,5$ for all $m \in \max(A)$, we simply choose $u = v = z_1$, $\sigma = \tau = \tau' = id_\psi$. Then in this case we must find g, h A, such that

1) $g^2 + ah^2 \in A^*$

2) $\psi(x') = g^2\psi(x) + 2agh\psi(x,y) + a^2h^2\psi(y) \in A^*$

3) $\psi(y') = h^2\psi(x) - 2gh\,\psi(x,y) + g^2\psi(y) \in A^*$.

Using the chinese remainder theorem for A and the fact $\psi(x) + a\psi(y) \in A^*$, we easily verify that such elements actually exist. If $|A/m| = 3$, 5 for some $m \in \max(A)$, then again using the chinese remainder theorem we can find $u, v \in \psi$, $g, h \in A$, such that $\psi(u)$, $\psi(v)$, $g^2\psi(u) + ah^2\psi(v)$, $\psi(x')$, $\psi(y') \in A^*$ are satisfied. We will omit the proof of this fact. Now applying part (i), we get a similarity $\Phi': \varphi \longrightarrow \varphi$ with norm $\varphi(z') = (g^2\psi(u) + ah^2\psi(v))\varphi(z)$, such that $\Phi'(z') = \varphi(z')z_1$. We define a new similarity $\Delta: \varphi \longrightarrow \varphi$ with norm $g^2\psi(u) + ah^2\psi(v)$

by

$$\Delta(r + t\otimes w) = g\sigma(r) + h\psi(v)a\tau'(w) + t\otimes[h\tau(r) - g\psi(u)\tau\sigma^{-1}\tau'(w)]$$

for all $r,w \in \psi$. Then $\Delta(z) = z'$. The similarity

$$\Delta' = (g^2\psi(u) + ah^2\psi(v))^{-1}\Delta$$

has norm $(g^2\psi(u) + ah^2\psi(v))^{-1}$, and therefore $\Phi = \Phi'\Delta'$ is a similarity of φ with norm $\varphi(z)$. On the other hand we have

$$\Phi(z) = \Phi'((g^2\psi(u) + ah^2\psi(v))^{-1}z') = \varphi(z)z_1 ,$$

that is Φ has all required properties. This proves the theorem.

(2.16) <u>Corollary</u>. (see $[K]_2$) Let A be a semi local ring with $|A/m| > 2$ for all $m \in \max(A)$. Then the bilinear Pfister space $\langle\langle a_1,\dots,a_n\rangle\rangle$ is round.

It would be desirable to find a proof of this result for all semi local rings.
Another way to express the content of theorem (2.8) is the following

(2.17) <u>Corollary</u>. With the same assumtions as in (2.8), the group of similarities $\Sigma(\varphi)$ of the Pfister space φ operates transitively on the set φ^* of strictly anisotropic elements of φ.

<u>Proof</u>. Let $z \in \varphi$ be strictly anisotropic and consider a typical basis $\{z_1,\dots,z_{2^n}\}$ of φ. Then it suffices to find $\sigma \in \Sigma(\varphi)$ such that $\sigma(z) = z_1$. According to (2.8) we can find $\sigma \in \Sigma(\varphi)$ with norm $\varphi(\varphi(z)^{-1}z) = \varphi(z)^{-1}$ and

$$\sigma(\varphi(z)^{-1}z) = \varphi(z)^{-1}z_1 ,$$

that is $\sigma(z) = z_1$. This proves our assertion.

In particular we get a cancellation theorem for one-dimensional subspaces of bilinear Pfister spaces, namely

(2.18) <u>Corollary</u>. Let φ be a bilinear Pfister space over the semi local ring A with $|A/m| > 2$ for all $m \in \max(A)$. For any $x,y \in \varphi$ with $\varphi(x) = \varphi(y) \in A^*$ there exists $\sigma \in O(\varphi)$ with $\sigma(x) = y$. In particular, if $\varphi = \langle a\rangle \perp \varphi_1 \cong \langle a\rangle \perp \varphi_2$ for some $a \in A^*$, then $\varphi_1 \cong \varphi_2$.

Now we are going to prove that some of the above results also hold for any semi local ring A, provided that we consider a more restricted class of bilinear Pfister spaces, namely, those Pfister spaces of the form $\varphi = \langle 1,1\rangle \otimes \langle 1,a_2\rangle \otimes \ldots \otimes \langle 1,a_n\rangle = 2 \times \psi$, where ψ is a (n-1)-fold Pfister space. In this case we have

(2.19) Corollary. Let ψ be any bilinear Pfister space over the semi local ring A. Then $\varphi = 2 \times \psi$ is round.

Proof. Let us consider again the cubic Frobenius extension $B = A[X]/(X^3 + 6X^2 - X + 1)$ with the trace map s, as defined in (2.3). Since $|B/M| \geq 7$ holds for all $M \in \max(B)$, we conclude from (2.8) that $B \otimes \varphi$ is round over B. Now let us take $\lambda \in \underline{D}(\varphi)^*$. Therefore $B \otimes \varphi \cong \langle\lambda\rangle \otimes B \otimes \varphi \cong B \otimes(\langle\lambda\rangle \otimes \varphi)$. Applying the transfer map s_* we get

$$\varphi \perp \begin{pmatrix} 0 & 1 \\ 1 & 6 \end{pmatrix} \otimes \varphi \cong \langle\lambda\rangle \otimes \varphi \perp \begin{pmatrix} 0 & 1 \\ 1 & 6 \end{pmatrix} \otimes \langle\lambda\rangle \otimes \varphi$$

But it is easy to see that

$$2A = V\left(\begin{pmatrix} 0 & 1 \\ 1 & 6 \end{pmatrix} \otimes \varphi \right) = V\left(\begin{pmatrix} 0 & 1 \\ 1 & 6 \end{pmatrix} \otimes \langle\lambda\rangle \otimes \varphi \right) \subseteq V(\varphi) ,$$

so that we can apply the lemma (4.6), chp.III, to conclude that

$$\varphi \perp m \times \begin{pmatrix} 0 & 1 \\ 1 & 0 \end{pmatrix} \cong \langle\lambda\rangle \otimes \varphi \perp m \times \begin{pmatrix} 0 & 1 \\ 1 & 0 \end{pmatrix}$$

where $m = \dim \varphi$. Since $2 \in \underline{D}(\varphi)$, we now use the cancellation theorem (4.5), chap.III, and we get

$$\varphi \cong \langle\lambda\rangle \otimes \varphi$$

that is φ is round. This proves the proposition.

In particular the Pfister space $2^n \times \langle 1\rangle = \langle\langle 1,\ldots,1\rangle\rangle$ (n-times) is round over any semi local ring A for all $n \geq 0$. Thus we have (see $[K]_3$ $[Pf]_{2,3}$) the following

(2.20) <u>Corollary</u>. For any integer $n \geq 0$ let $Q(2^n)$ be the set of units of A, which are sums of 2^n squares in A. Then $Q(2^n)$ is a subgroup of A^*.

Of course we can use (2.4) to deduce a similar result as in (2.20) for the quadratic case. To this end let us choose an entire $h \in \mathbb{Z}$ with the property $1 - 4h \in A^*$. Using the fact that A is semi local, we can find infinitely many integers with this property. Now we define the quadratic Pfister space $q = 2^n \times [1,h] = \langle\langle 1,\ldots,1,h]]$ and denote the set of units of A of the form

$$\sum_{i=1}^{n} (\alpha_i^2 + \alpha_i\beta_i + h\beta_i^2)$$

by $D(n,h)$, that is $D(n,h) = \underline{D}(q)^*$. Since q is round, we conclude that $D(n,h)$ is a subgroup of A^* for all $n \geq 0$ and all $h \in \mathbb{Z}$ with $1 - 4h \in A$. If $2 \in A^*$, then this result follows immediately from (2.20).

Our purpose now is to extend (2.18) to bilinear Pfister spaces of the form $\varphi = 2 \times \psi$ over any semi local ring A. Thus let us assume that

$$\varphi = \langle a\rangle \perp \varphi_1 = \langle a\rangle \perp \varphi_2$$

for some $a \in A^*$. Since φ is round (see(2.19)), it follows that $\langle a\rangle \otimes \varphi \cong \varphi$.Now we may scale with $\langle a\rangle$ to reduce the problem to the case $a = 1$. On the other hand we have $\varphi = \psi \perp \psi = \langle 1\rangle \perp \psi' \perp \psi$, thus we can assume that $\varphi_2 = \psi' \perp \psi$. We show that $\varphi_1 \cong \psi' \perp \psi$. Considering again the cubic extension B, which was defined in (2.3), we deduce from (2.18) that

$$B \otimes \varphi_1 \cong B \otimes(\psi' \perp \psi),$$

and therefore using the transfer map s_* we get

$$\varphi_1 \perp \begin{pmatrix} 0 & 1 \\ 1 & 6 \end{pmatrix} \otimes \varphi_1 \cong \psi' \perp \psi \perp \begin{pmatrix} 0 & 1 \\ 1 & 6 \end{pmatrix} \otimes(\psi' \perp \psi)$$

The lemma (4.6), chap.III, now implies

$$\varphi_1 \perp m \times \begin{pmatrix} 0 & 1 \\ 1 & 0 \end{pmatrix} \cong \psi' \perp \psi \perp m \times \begin{pmatrix} 0 & 1 \\ 1 & 0 \end{pmatrix}$$

($m = \dim \varphi_1$). But if $\dim \psi > 1$, then for any $b \in \underline{D}(\psi')^*$ we have $2b \in$

$\underline{D}(\psi' \perp \psi)$, thus we may use the cancellation theorem (4.5), chap.III, to conclude that

$$\varphi_1 \cong \psi' \perp \psi$$

If dim $\psi = 1$, i.e. $\psi = \langle 1\rangle$, we compare the determinants of $\langle 1\rangle \perp \varphi_1$ and $\langle 1\rangle \perp \langle 1\rangle$, and we get $\varphi_1 \cong \langle 1\rangle$. This finish the proof of the following.

(2.21) Proposition. Let ψ be a bilinear Pfister space over the semi local ring A and $\varphi = 2 \times \psi$. Then for any $x,y \in \varphi$ with $\varphi(x) = \varphi(y) \in A^*$ there exists $\sigma \in O(\varphi)$ with $\sigma(x) = y$.

(2.22) Corollary. Let A,φ be as in (2.21). Then for any $z \in \varphi$ with $\varphi(z) \in A^*$ there exists a similarity σ of φ with norm $\varphi(z)$, such that

$$\sigma(z) = \varphi(z) z_1 ,$$

where $\{z_1, \ldots z_{2^n}\}$ is a typical basis of φ.

Proof. According to (2.19) we can find a similarity $\sigma' : \varphi \longrightarrow \varphi$ with norm $\varphi(z)$. In particular we have $\varphi(\sigma'(z)) = \varphi(z)^2$. Therefore $\varphi(\varphi(z)z_1) = \varphi(z)^2 = \varphi(\sigma'(z))$. By (2.21), we conclude that there exists an automorphism τ of φ such that $\tau(\sigma'(z)) = \varphi(z)z_1$. Thus, $\sigma = \tau \cdot \sigma'$ does the job.

In particular for any $\omega \in \varphi$ we have, under the assumptions of (2.22),

$$\varphi(z)\varphi(\omega) = (z,\omega)^2 + \varphi(\omega')$$

with $\omega' \in \varphi' = \langle z_1\rangle^{\perp}$. Here the coefficients of ω' depend linearly and rationally on the coefficients of ω and z, respectively. For example, let us consider the Pfister space $\varphi = 2^n \times \langle 1\rangle$ over the semi local ring A. Take elements $u_1, \ldots, u_{2^n}$, $v_1, \ldots, v_{2^n} \in A$, such that $u_1^2 + \ldots + u_{2^n}^2 \in A^*$. Then there are elements $\omega_2, \ldots, \omega_{2^n}$ depending linearly on $v_1, \ldots, v_{2^n}$ and rationally on $u_1, \ldots, u_{2^n}$, such that

$$(2.23) \qquad (u_1^2 + \ldots + u_{2^n}^2)(v_1^2 + \ldots + v_{2^n}^2) = (u_1v_1 + \ldots + u_{2^n}v_{2^n})^2 + \omega_2^2 + \ldots + \omega_{2^n}^2$$

(see [Pf]$_1$ for the field case).

§ 3. Isotropic Pfister spaces.

Let A be a semi local ring. One of the most basic results in the Pfister theory of quadratic forms is the following theorem.

(3.1) Theorem. Let (E,q) be a quadratic space over A of the form $E = <1,a> \otimes F$, where (F,q) is a round quadratic space and $a \in A^*$. If E is isotropic, then

$$-a \in \underline{D}(F)^*$$

and in particular, $(E,q) \cong \mathbb{H}[F]$ is hyperbolic.

In section 5 of chapter III we have proved a more general version of this result (see (5.3)). Now we want to give a slightly different proof of this fact. If dim F = 1, that is $F = <1>$, we have $E = <1,a>$ and consequently the assumption of (3.1) implies $x^2 + ay^2 = 0$, where both x,y are units. Hence $-a \in A^{*2}$. Suppose now dim $F \geq 2$ and $E = F \perp <t> \otimes F$ with $(t,t) = a$. Let $z = x + t \otimes y \in E$ be isotropic, that is $q(z) = q(x) + aq(y) = 0$ and z primitive. Using (5.2) (iii), chap. III, we can find $\sigma \in O^+(E)$ such that $\sigma(z) = x' + t \otimes y'$ and $q(x')$, $q(y') \in A^*$. Thus we may assume that $q(x), q(y) \in A^*$. From $q(x) + aq(y) = 0$ it follows that

$$-a = q(y)^{-1}q(x) \in \underline{D}(F)^* ,$$

since F is round. This proves (3.1) again.

(3.2) Corollary. Let $E = <<a_1,\ldots,a_n,b]]$ be a quadratic Pfister space over A. If E is isotropic, then E is hyperbolic.

Proof. For $E = [1,b]$ there is nothing to prove. For $n \geq 1$ we write $E = <1,a_1> \otimes F$ with $F = <<a_2,\ldots,a_n,b]]$ (if n = 1, then $F = [1,b]$). Since F is round, the result follows.

(3.3) Remark. The corollary (3.2) is not true for bilinear Pfister spaces. For example, let us consider the local ring $A = \mathbb{Z}/4\mathbb{Z}$ and the Pfister space $\varphi = 4 \times <1> = <<1,1>>$ over A. Clearly φ is isotropic, since 4 = 0. Let $\{z_1,z_2,z_3,z_4\}$ be the corresponding typical basis for φ. Considering the following basis $\{z_4, z_1+z_2+z_3, z_1-z_3, z_2-z_4\}$ for φ, we get the isomorphism

$$<1,1,1,1> \cong <1,-1> \perp \begin{pmatrix} 2 & 1 \\ 1 & 2 \end{pmatrix}$$

But it can easily be seen that $\begin{pmatrix} 2 & 1 \\ 1 & 2 \end{pmatrix}$ is anisotropic. This implies that $4 \times <1>$ cannot be metabolic, since over $\mathbb{Z}/4\mathbb{Z}$ the anisotropic part of a bilinear space is uniquely determined (see § 8 in $[K]_1$). In any case we have the following weaker result for bilinear Pfister spaces.

(3.4) Proposition. Let $\varphi = <<a_1,\dots,a_n>> = <1> \perp \varphi'$ be a bilinear Pfister space over A, which is isotropic. Assume one of the following conditions:

i) $|A/m| \geq 3$ for all $m \in \max(A)$

ii) $a_1 = 1$, i.e. $\varphi = 2 \times \psi$ with $\psi = <<a_2,\dots,a_n>>$.

Then φ' represents -1 and in particular $2 \times \varphi \cong M(\varphi)$ is metabolic.

Proof. Choose $x \in \varphi$ isotropic, i.e. x primitive and $\varphi(x) = 0$. We set $\varphi = <z_1> \perp \varphi'$ with $\varphi' = <z_2,\dots,z_{2^n}>$ and $\varphi(z_1) = 1$. Then $x = \alpha z_1 + u$, where $\alpha \in A$, $u \in \varphi'$ satisfy the relation $\alpha^2 + \varphi(u) = 0$. Of course if $\alpha \in A^*$, then $\varphi(\alpha^{-1}u) = -1$, and hence there is nothing to prove. Let us now consider the general case. We claim:

(*) There exists $y \in \varphi$ with $\varphi(y)$, $(y,x) \in A^*$. To prove (*) we may assume that A is a field. Because if $\bar{y} \in \varphi(m)$ can be found with $\varphi(\bar{y})$, $(\bar{y},\bar{x}) \neq 0$ in A/m for all $m \in \max(A)$, we just lift these elements to $y \in \varphi$, which obviously satisfies (*). Now suppose that A is a field. If $\alpha \neq 0$, we simply take $y = z_1$. Assume $\alpha = 0$, i.e. $x = u \in \varphi'$, $u \neq 0$. Then $u = \alpha_2 z_2 + \dots + \alpha_{2^n} z_{2^n}$ and some $\alpha_i \neq 0$. In this case we take $y = z_i$.
Let us return to the semi local case. With $y \in \varphi$ as in (*), we get from (2.8) in case (i) and from (2.22) in case (ii), that

$$0 = \varphi(y)\varphi(x) = (y,x)^2 + \varphi(\omega)$$

with $\omega \in \varphi'$. Since $(y,x) \in A^*$, we obtain $-1 = \varphi((y,x)^{-1}\omega) \in \underline{D}(\varphi')$, proving the first assertion. Since φ is round (use (2.8) or (2.22)), it follows that $2 \times \varphi = \varphi \perp \varphi \cong \varphi \perp <-1> \otimes \varphi = M(\varphi)$. Thus, the proof of the proposition is complete.

Without the assumptions (i) or (ii) of (3.4), we can only show the following weaker result:

(3.5) Proposition. Let φ be a bilinear Pfister space over the semi local ring A. If φ is isotropic, then $2 \times [\varphi] = 0$ in W(A).

Proof. Let $B = A[X]/(X^3 + 6X^2 - X + 1)$ be the cubic Frobenius extension of A, which was defined in (2.3), with the corresponding trace map s. Since $\varphi \otimes B$ is isotropic over B, we deduce from (3.4) (i) that $2 \times \varphi \otimes B$ is metabolic, and therefore $2 \times [\varphi \otimes B] = 0$ in W(B). Applying the transfer map s_*, we get

$$2 \times s_*[\varphi \otimes B] = 2 \times [\varphi]s_*(<1>) = 0 ,$$

and since $s_*(<1>) \sim <1>$, we conclude that $2 \times [\varphi] = 0$ in W(A).

§ 4. Further results on quadratic Pfister spaces.

In this section we will consider throughout a semi local ring A with $|A/m| \geq 3$ for all $m \in \max(A)$. Our purpose is to extend some results of $[E\text{-}L]_3$ to such class of rings.

(4.1) Lemma. Let $q = <<a_1,\dots,a_n,a]]$ be a quadratic Pfister space over A. If q contains the space [1,b], then

$$q \cong <<b_1,\dots,b_n,b]]$$

for some $b_1,\dots,b_n \in A^*$.

Proof. If $n = 0$, i.e. $q = [1,a]$, then there is nothing to prove. If $n = 1$ and $q = <1,a_1> \otimes [1,a] \cong [1,b] \perp <b_1> \otimes [1,d]$ for some $b_1 \in A^*$, $d \in A$, we deduce by comparing the Arf-invariants that $A(\wp^{-1}(-b)) \cong A(\wp^{-1}(-d))$, that is $[1,b] \cong [1,d]$. Hence $q \cong <1,b_1> \otimes [1,b]$. Let us now assume that $n > 1$. We proceed by induction on n. We set $q_1 = <<a_2,\dots,a_n,a]]$. Therefore $q = <1,a_1> \otimes q_1 = q_1 \perp <a_1> \otimes q_1$. If $\{z_1,\dots,z_{2^{n-1}}\}$ is a typical basis for $<<a_2,\dots,a_n>>$ and $\{e,f\}$ is a basis of [1,a] with $q(e) = 1$, $(e,f) = 1$, $q(f) = a$, then $\{z_1 \otimes e,\ z_1 \otimes f,\ \dots,\ z_{2^{n-1}} \otimes e,\ z_{2^{n-1}} \otimes f\}$ is a typical basis for q_1. Note that $q(z_1 \otimes e) = 1$. On the other hand we have $[1,b] = <g,h> \subset q$, where $q(g) = (g,h) = 1$, $q(h) = b$. According to (4.1), chap. III, one can find $\sigma \in O(q)$ such that $\sigma(g) = z_1 \otimes e$. Since $<g,h> \cong <\sigma(g), \sigma(h)>$, we can replace $<g,h>$ by $<\sigma(g), \sigma(h)>$, that is, we may assume without restriction that $g = z_1 \otimes e$. Now, if $h = x + t \otimes y$, where $x,y \in q_1$ and $\langle a_1 \rangle = At$ with $(t,t) = a_1$, one can apply the method used in the proof of (5.1) and (5.2), chap. III, to assure that $q_1(y)$, $1 - 4q_1(x) \in A^*$. This can be achieved without disturbing the assumption $g = z_1 \otimes e$. Now $b = q(h) = q_1(x) + a_1 q_1(y)$ and $(z_1 \otimes e,h) = (z_1 \otimes e,x) = 1$. The relation $1 - 4q_1(x) \in A^*$ implies that $[1,q_1(x)] = \langle z_1 \otimes e,x \rangle$ is a subspace of q_1. Hence we use the induction hypothese to conclude that

$$q_1 \cong \langle\langle b_2,\dots,b_n,q_1(x)]]$$

for some $b_2,\dots,b_n \in A^*$. But q_1 is round and $q_1(y) \in A^*$, so that $\langle q_1(y) \rangle \otimes q_1 \cong q_1$. Therefore

$$\begin{aligned} q &\cong \langle 1,a_1q_1(y) \rangle \otimes q_1 \\ &\cong \langle 1,a_1q_1(y) \rangle \otimes \langle\langle b_2,\dots,b_n,q_1(x)]] \\ &\cong \langle\langle b_2,\dots,b_n\rangle\rangle \otimes \langle\langle a_1q_1(y),q_1(x)]] \end{aligned}$$

But the space $[1,b] = [1,q_1(x) + a_1q_1(y)]$ is contained in $\langle\langle a_1q_1(y), q_1(x)]]$, so that $\langle\langle a_1q_1(y), q_1(x)]] \cong \langle\langle b_1,b]]$ with a suitable $b_1 \in A^*$. Inserting this last relation in the right side of the above isomorphism, it follows that $q \cong \langle\langle b_1,\dots,b_n,b]]$, proving the lemma.

(4.2) Remarks. 1) Using (4.1) we can prove (3.2) again. Because if $q = \langle\langle a_1,\dots,a_n,a]]$ is isotropic, then q contains the subspace $[0,0] \cong [1,0]$, and therefore $q \cong \langle\langle b_1,\dots,b_n,0]] \cong 2^n \times \mathbb{H}$.
2) Let us assume $2 \in A^*$. If $\varphi = \langle\langle a_1,\dots,a_n\rangle\rangle = \langle 1 \rangle \perp \varphi'$ is a Pfister space over A and $b \in \underline{D}(\varphi')^*$, then $\langle 1,b \rangle$ is a subspace of φ. Thus we can use (4.1) to conclude that $\varphi \cong \langle\langle b,b_2,\dots,b_n\rangle\rangle$ with suitable $b_2,\dots,b_n \in A^*$ (see $[E-L]_3$ for the field case).

Our purpose is to generalize (4.1). To this end we need the following result (see $[E-L]_3$).

(4.3) Lemma. Let $\rho = \langle\langle b_1,\dots,b_n\rangle\rangle$ be a bilinear Pfister space and $q = \langle\langle a_1,\dots,a_m,a]]$ be a quadratic Pfister space over A. Then for every $b \in \underline{D}(\rho' \otimes q)^*$ there exist $c_2,\dots,c_n \in A^*$, such that

$$\rho \otimes q \cong \langle\langle b,c_2,\dots,c_n\rangle\rangle \otimes q$$

<u>Proof</u>. If $n = 1$, then $\rho' = \langle b_1 \rangle$ and hence $b = b_1 c$ with some $c \in \underline{D}(q)^*$. Hence $\rho \otimes q \cong q \perp \langle b_1 \rangle \otimes q \cong q \perp \langle b_1 c \rangle \otimes q = \langle 1,b \rangle \otimes q$, since q is round. Assume now $n > 1$. We proceed by induction on n. We set $\rho = \langle 1,b_1 \rangle \otimes \rho_1$ with $\rho_1 = \langle\langle b_2,\ldots,b_n \rangle\rangle$. Using the cancellation theorem (4.3), chap. III, we get $\rho' \otimes q \cong \rho_1' \otimes q \perp \langle b_1 \rangle \otimes \rho_1 \otimes q$, and consequently $b = d_1 + b_1 d_2$ with $d_1 \in \underline{D}(\rho_1' \otimes q)^*$, $d_2 \in \underline{D}(\rho_1 \otimes q)^*$ (use (5.2), chap. III). But $\rho_1 \otimes q$ is round, so that

$$\rho \otimes q = \langle 1,b_1 \rangle \otimes \rho_1 \otimes q \cong \langle 1,b_1 d_2 \rangle \otimes \rho_1 \otimes q .$$

On the other hand, we use the induction hypothesis to conclude that $\rho_1 \otimes q \cong \langle\langle d_1,c_3,\ldots,c_n \rangle\rangle \otimes q$ with suitable $c_i \in A^*$ (if $n = 2$, then no c_i appears). Therefore

$$\rho \otimes q \cong \langle 1,b_1 d_2 \rangle \otimes \langle 1,d_1 \rangle \otimes \langle\langle c_3,\ldots,c_n \rangle\rangle \otimes q$$

But $\langle\langle b_1 d_2,d_1 \rangle\rangle \cong \langle\langle d_1 + b_1 d_2,c_2 \rangle\rangle$ with some $c_2 \in A^*$, thus it follows that

$$\rho \otimes q \cong \langle\langle b,c_2,\ldots,c_n \rangle\rangle \otimes q$$

proving our assertion.

Now we present the announced generalization of (4.1).

(4.4) <u>Theorem</u>. Let p and q be two quadratic Pfister space over A with $q \cong p \perp q_o$ with some q_o. Then there exists a bilinear Pfister space τ such that

$$q \cong \tau \otimes p$$

<u>Proof</u>. Since the case $p = [1,b]$ has been considered in (4.1), we may assume that $p = \langle\langle b_1,\ldots,b_r,b]]$ with $r \geq 1$. We proceed by induction on r. To this end let us set $p = \langle 1,b_1 \rangle \otimes p_1$ with $p_1 = \langle\langle b_2,\ldots,b_r,b]]$ (of course $p_1 = [1,b]$ if $r = 1$). Hence $q \cong p_1 \perp \langle b_1 \rangle \otimes p_1 \perp q_o$ with a suitable q_o. Using the induction hypothesis we deduce $q \cong \tau_1 \otimes p_1$ with a suitable bilinear Pfister space τ_1. Hence

$$p_1 \perp \tau_1' \otimes p_1 \cong p_1 \perp \langle b_1 \rangle \otimes p_1 \perp q_o$$

and therefore (see (4.3), chap. III)

$$\tau_1' \otimes p_1 \cong \langle b_1\rangle \otimes p_1 \perp q_o \, .$$

This implies $b_1 \in \underline{D}(\tau_1' \otimes p_1)$, thus using (4.3) we can find $c_2, \ldots, c_e \in A^*$ such that

$$\begin{aligned} q \cong \tau_1 \otimes p_1 &\cong \langle\langle b_1, c_2, \ldots, c_e\rangle\rangle \otimes p_1 \\ &\cong \langle\langle c_2, \ldots, c_e\rangle\rangle \otimes \langle 1, b_1\rangle \otimes p_1 \\ &\cong \langle\langle c_2, \ldots, c_e\rangle\rangle \otimes p \end{aligned}$$

i.e. $\tau = \langle\langle c_2, \ldots, c_e\rangle\rangle$ does the job. This proves the theorem.
The results above deal with the devisibility of quadratic Pfister spaces by Pfister spaces. Thus we are led to the following question (see [B-K], [Sh-W]): let φ, ψ be two bilinear spaces and q a quadratic Pfister space over A, such that $\varphi \otimes q = \psi \otimes q$. Look for another equivalent relations between φ, ψ and q. In chapter V, section 8 we shall consider a related question.
For simplicity let us assume that $|A/m| \geq 5$ for all $m \in \max(A)$. Given two bilinear spaces $\varphi = \langle a_1, \ldots, a_n\rangle$ and $\psi = \langle b_1, \ldots, b_n\rangle$ ($a_i, b_i \in A^*$), we say that φ and ψ are Witt-equivalent, if there exist indices $1 \leq i, j \leq n$ with $\langle a_i, a_j\rangle \cong \langle b_i, b_j\rangle$ and $a_k = b_k$ for all $k \neq i,j$. Let q be a round quadratic space over A. Following [Sh-W] we call φ and ψ simply q-equivalent, if there exists one index $1 \leq i \leq n$ such that $a_i = \lambda b_i$ for some $\lambda \in N(q)$ and $a_k = b_k$ for all $k \neq i$. For the Witt-equivalence we use the symbol $\sim$ and for the symply q-equivalence the symbol $\underset{q}{\sim}$. Combining these two equivalences we get the q-equivalence, which will be denoted by $\underset{q}{\approx}$. It is obvious that $\varphi \underset{q}{\approx} \psi$ implies $\varphi \otimes q \cong \psi \otimes q$. Now we are going to show that the converse is also true. Thus let us assume that $\varphi \otimes q = \psi \otimes q$. Since $1 \in \underline{D}(q)$ (because q is round), we get $b_1 \in \underline{D}(\varphi \otimes q)^*$. Now $\varphi \otimes q = \langle a_1\rangle \otimes q \perp \ldots \perp \langle a_n\rangle \otimes q$. Applying the method used in the proof of (5.1) and (5.2), chap. III, we can find elements $x_1, \ldots, x_n \in q$ such that

a) $q(x_i) \in A^*$ for $1 \leq i \leq n$

b) $a_1 q(x_1) + \ldots + a_r q(x_r) \in A^*$ for $1 \leq r \leq n$

c) $b_1 = a_1 q(x_1) + \ldots + a_n q(x_n)$

(at this place we need the assumption $|A/m| \geq 5$ for all $m \in \max(A)$).
Using these properties we get inductively

$$\varphi \underset{q}{\sim} \langle a_1 q(x_1),\dots,a_n\rangle \underset{q}{\sim} \cdots \underset{q}{\sim} \langle a_1 q(x_1),\dots,a_n q(x_n)\rangle$$

and on the other hand, according to (b),

$$\langle a_1 q(x_1),\dots,a_n q(x_n)\rangle \sim \langle a_1 q(x_1) + a_2 q(x_2), c_2,\dots,a_n q(x_n)\rangle$$
$$\dots \sim \langle a_1 q(x_1) + \dots + a_n q(x_n), c_2,\dots c_n\rangle ,$$

that is $\varphi \underset{q}{\approx} \langle b_1,c_2,\dots,c_n\rangle$ with suitable $c_2,\dots,c_n \in A^*$. This relation implies $\varphi \otimes q \cong \langle b_1,c_2,\dots,c_n\rangle \otimes q$ and in virtue of $\varphi \otimes q \cong \langle b_1,b_2\dots,b_n\rangle \otimes q$, we get $\langle b_1,b_2,\dots,b_n\rangle \otimes q \cong \langle b_1,c_2,\dots,c_n\rangle \otimes q$. Cancelling $\langle b_1\rangle \otimes q$, we get $\langle b_2,\dots,b_n\rangle \otimes q \cong \langle c_2,\dots,c_n\rangle \otimes q$. Now use induction to conclude that $\langle c_2,\dots, c_n\rangle \underset{q}{\approx} \langle b_2,\dots b_n\rangle$. This implies $\varphi \underset{q}{\approx} \psi$. Thus we have shown.

(4.5) <u>Proposition</u>. Under the above assumptions the following assertions are equivalent

i) $\varphi \otimes q \cong \psi \otimes q$
ii) $\varphi \underset{q}{\approx} \psi$

Now let us denote the annulator of the class $[q] \in Wq(A)$ in $W(A)$ by $\mathrm{Ann}(q)$, that is $\mathrm{Ann}(q) = \{[\rho] \in W(A) \mid \rho \otimes q \sim 0\}$. Using the cancelation theorem (4.3), chap. III, we easily see that the following assertions are equivalent

i) $\varphi \otimes q = \psi \otimes q$
iii) $[\varphi] - [\psi] \in \mathrm{Ann}(q)$ and $\dim \varphi = \dim \psi$.
If q is round, so they are also equivalent with (ii) $\varphi \underset{q}{\approx} \psi$ (see (4.5))

We use the equivalence (ii) $\Longleftrightarrow$ (iii) to compute Ann(q) for a quadratic round space q (see $[K]_3$, and compare with § 8, chap. V).

(4.6) <u>Theorem</u>. Let q be a quadratic round space over the semi local ring A with $|A/m| \geq 5$ for all $m \in \max(A)$. Then

$$\mathrm{Ann}(q) = \sum_{\lambda \in N(q)} W(A)\ \langle 1,-\lambda\rangle$$

(4.7) <u>Remark</u>. No answer is known for $\mathrm{Ann}(\varphi) = \{[\psi] \in W(A) \mid \psi \otimes \varphi \sim 0\}$, where φ is a bilinear Pfister form over a semi local ring A with $2 \notin A$, not even in the case $\varphi = \langle 1,a\rangle$, $a \in A^*$.

CHAPTER V

Structure of Witt rings

§ 1. Introduction.

Let A be a commutative ring with 1. In this chapter Wq(A) denotes throughout the Witt ring of quadratic spaces of constant rank over A. Correspondingly, W(A) is the Witt ring of bilinear spaces of constant rank over A.
The map, which associates with every $[E] \in Wq(A)$ the class rank (E) (mod $2\mathbb{Z}$) $\in \mathbb{Z}/2\mathbb{Z}$, is a ring homomorphism

(1.1) $$d : Wq(A) \longrightarrow \mathbb{Z}/2\mathbb{Z}.$$

Correspondingly, we have a ring homomorphism

(1.2) $$d : W(A) \longrightarrow \mathbb{Z}/2\mathbb{Z}.$$

Let us denote the kernels of these homomorphisms by $Wq(A)_o$ and $I_A = W(A)_o$, respectively. If $2 \notin A^*$, then $Wq(A) = Wq(A)_o$ because in this case every quadratic space has even rank. If $2 \in A^*$, then $Wq(A) = W(A)$ and $Wq(A)_o = I_A$. In this case it holds $W(A)/I_A \cong \mathbb{Z}/2\mathbb{Z}$.

(1.3) Remark. Let A be a semi local ring. In virtue of (3.4) and (3.5), chap. I we see that $Wq(A)_o$ is additively generated by the classes of the spaces $\langle a\rangle \otimes [1,b]$, $(a, 1-4b \in A^*)$, while I_A is additively generated by the classes $[\langle 1,a\rangle]$, $a \in A^*$.

Now we consider some simple examples. To this end we need the following result.

(1.4) Lemma. Let A be a semi local ring, which is complete in the r-adic topology (r = Jacobson radical of A). Let (E,q) be a quadratic space over A. Then if (E(m),q(m)) is isotropic over A/m for all m $\in$ max(A), then (E,q) is isotropic over A.

Proof. Assume that (E(m),q(m)) is isotropic over A/m for all m $\in$ max(A). Let {e(m),f(m)} be a hyperbolic pair in E(m). Using the chinese remainder theorem we can find $e,f \in E$, whose residue classes modulo $m \in \max(A)$ are $e(m), f(m)$, respectively. We set $a = q(e)$, $b = q(f)$ and $c = (e,f)$. Hence $a,b \in r$ and $c \in A^*$. Now, the reduction modulo r of the polynomial $f(X) = aX^2 + cX + b$ is $\bar{f}(X) = X$, thus $\bar{f}'(X) = 1$. Therefore, using the well-known method of approximation to the roots of equations of Newton, we can find $\lambda \in A$, such that $f(\lambda) = 0$ (here we use the completeness of A). Hence the element $g = \lambda e + f$ is primitive and isotropic, that is (E,q) is isotropic.

(1.5) Corollary. Let A be a complete semi local ring in the r-adic topology. Then the canonical homomorphism $A \longrightarrow A/r$ induces an isomorphism

$$Wq(A) \xrightarrow{\sim} Wq(A/r)$$

Proof. From (1.4) it follows immediately that $Wq(A) \longrightarrow Wq(A/r)$ is one to one. On the other hand, since every space $(\bar{E},\bar{q})$ over A/r of the form $<\bar{a}> \otimes [1,\bar{b}]$ can be lifted to a space $<a> \otimes [1,b]$ over A, we conclude from (1.3), that the above homomorphism is onto. This proves the corollary.

Let us suppose now that A is a finite local ring with maximal ideal m. In particular A/m is a finite field, say $A/m = \mathbb{F}_{p^s}$ for some prime number p and integer $s \geq 1$. This prime number is the unique prime with the property $p \in m$ and s is determined as the smallest positive integer with the property $a^{p^s} - a \in m$ for all $a \in A$. Since A is m-adically complete, we conclude from (1.5) that $Wq(A) \xrightarrow{\sim} Wq(\mathbb{F}_{p^s})$. Now if p is odd, we have (see [L])

$$(1.6) \qquad Wq(\mathbb{F}_{p^s}) = \begin{cases} \mathbb{Z}/2\mathbb{Z}\,[\mathbb{F}_{p^s}^*/\mathbb{F}_{p^s}^{*2}] & \text{for } p^s \equiv 1 \pmod 4 \\ \mathbb{Z}/4\mathbb{Z} & \text{for } p^s \equiv 3 \pmod 4 \end{cases}$$

If $p = 2$, we set $k = \mathbb{F}_{2^s}$. Since k is perfect, we have $<a> \cong <1>$ for all $a \in k^*$, and therefore Wq(k) is generated by the spaces $[1,b]$, $b \in k$. Since $[1,b_1] \perp [1,b_2] \cong \mathbb{H} \perp [1,b_1 + b_2]$ holds for all $b_1,b_2 \in k$, we see that Wq(k) consists of elements of the form $[[1,b]]$, $b \in k$. Now using the Arf-invariant we conclude that

$$a : Wq(k) \xrightarrow{\sim} k/\wp(k)$$

(this result is of course true for any perfect field of characteristic

2). If $k = \mathbb{F}_{2^s}$, then $k/\wp(k) \cong \mathbb{Z}/2\mathbb{Z}$, thus it follows

$$\text{(1.7)} \qquad Wq(\mathbb{F}_{2^s}) \cong \mathbb{Z}/2\mathbb{Z} .$$

For example, if $A = \mathbb{Z}/n\mathbb{Z}$ and $n = p_1^{a_1} \dots p_r^{a_r}$ is the prime decomposition of n, then we have $A \cong \mathbb{Z}/p_1^{a_1}\mathbb{Z} \times \dots \times \mathbb{Z}/p_r^{a_r}\mathbb{Z}$ and consequently

$$Wq(A) \cong Wq(\mathbb{F}_{p^1}) \times \dots \times Wq(\mathbb{F}_{p^r}) .$$

Since the rings $W(\mathbb{F}_{p^i})$ can be computed according to (1.6) and (1.7), we can compute $Wq(\mathbb{Z}/n\mathbb{Z})$.

§ 2. The discriminant map.

Let A be a semi local ring. For any quadratic space (E,q) over A we have defined the Arf-invariant $a(E) = [D(E)] \in \Delta(A)$, where $\Delta(A)$ is the group of quadratic separable extensions of A (see chap. II, § 1, 3). Since $a(\mathbb{H}[P]) = 1$, we get a map

$$\text{(2.1)} \qquad a : Wq(A) \longrightarrow \Delta(A),$$

the discriminant map. Since we have shown that $a(E_1 \perp E_2) = a(E_1)a(E_2)$ if rank (E_1) rank $(E_2) \equiv 0 \pmod 2$, it follows that a induces a group homomorphism

$$\text{(2.2)} \qquad a : Wq(A)_o \longrightarrow \Delta(A)$$

It should be noted that in case $2 \notin A^*$, then $Wq(A) = Wq(A)_o$ (since we are considering only free spaces) and in consequence (2.1) coincides with (2.2). If $2 \in A^*$, we shall identify $\Delta(A)$ with A^*/A^{*2} by the isomorphism given by $[A(\wp^{-1}(b))] \longrightarrow (1 + 4b) \bmod A^{*2}$. In this case let us write $(B) \in A^*/A^{*2}$ instead of $[B] \in \Delta(A)$. Now, using this notation we define the following group:

i) if $2 \notin A^*$ we define $Q(A) = \Delta(A)$

ii) if $2 \in A^*$ we define $Q(A) = \mathbb{Z}/2\mathbb{Z} \times \Delta(A)$, where the product on $Q(A)$ is given by

$$(\bar{n}_1,(B_1))\cdot(\bar{n}_2,(B_2)) = (\overline{n_1 + n_2}, (-1)^{n_1 n_2}(B_1 \circ B_2))$$

It is easy to see that

(2.3) Lemma. $Q(A)$ is an abelian group of exponent 4.

Using this group we now define the following map:

$$(2.4) \qquad \hat{a} : Wq(A) \longrightarrow Q(A)$$

by

$$\hat{a}([E]) = \begin{cases} [D(E)] & \text{if } 2 \notin A^* \\ (\overline{\dim (E)}, (D(E))) & \text{if } 2 \in A^* \end{cases}$$

for all $[E] \in Wq(A)$.
Then we have

(2.5) Theorem. i) $\hat{a}$ is a group epimorphism
ii) $\operatorname{Ker}(\hat{a}) = I\,Wq(A)_o$, that is

$$Wq(A)/I\,Wq(A)_o \xrightarrow{\sim} Q(A)$$

Proof. i) Let us first remark that if $2 \notin A^*$ and $[B] \in \Delta(A) = Q(A)$, then $\hat{a}([B]) = [B]$. If $2 \in A^*$ and $(d) \in A^*/A^{*2}$, we define the quadratic separable extension $B = A(\sqrt{d})$. Then it follows that $\hat{a}([B]) = (\bar{o},(d))$ and $\hat{a}([\frac{1}{2}d]) = (\bar{1},(d))$ in $Q(A)$. Thus, as soon as we have proved that a is a homomorphism, the above remarks shall imply that $\hat{a}$ is onto. If $2 \notin A^*$, then this follows immediately from (3.43), chap. III (see (2.2)). Thus we may assume $2 \in A^*$. Let E,F be two quadratic spaces over A and consider the orthogonal decompositions $E = \langle a_1\rangle \perp \ldots \perp \langle a_n\rangle$, $F = \langle b_1\rangle \perp \ldots \perp \langle b_m\rangle$ with $a_i, b_j \in A^*$. In virtue of the (4.5), chap. III we have

$$D(E) = A(\sqrt{(-1)^{\frac{n(n-1)}{2}} a_1 \ldots a_n})$$

$$D(F) = A(\sqrt{(-1)^{\frac{m(m-1)}{2}} b_1 \ldots b_m})$$

and hence $\hat{a}([E]) = (\bar{n}, ((-1)^{\frac{n(n-1)}{2}} a_1 \ldots a_n))$,
$\hat{a}([F]) = (\bar{m}, ((-1)^{\frac{m(m-1)}{2}} b_1 \ldots b_m)) \in Q(A)$. Correspondingly, we get

$$\hat{a}([E] \perp [F]) = (\overline{n+m}, ((-1)^{\frac{(n+m)(n+m-1)}{2}} a_1 \ldots a_n b_1 \ldots b_m)).$$

Now a straightforward computation shows that $\hat{a}([E] \perp [F]) =$

$\hat{a}(E)\cdot\hat{a}(F)$, proving part (i).
ii) Obviously $IWq(A)_o \subseteq Ker(\hat{a})$, because $IWq(A)_o$ is additively generated by the elements $\langle\langle d,b]]$ (with $d, 1-4b \in A^*$), which clearly lie in $Ker(\hat{a})$. Conversely, let us take $[E] \in Ker(\hat{a})$. In particular $\dim E$ is even and hence

$$E = \langle a_1\rangle \otimes [1,b_1] \perp \ldots \perp \langle a_n\rangle \otimes [1,b_n]$$

For $n = 1, 2$, one easily sees that $[E] \in IWq(A)_o$. Namely, if $n = 1$ we get $a([1,b_1]) = 1$, that is $[E] = 0$, and if $n = 2$ we conclude $a([1,b_1]) = a([1,b_2])$, i.e. $[1,b_1] \cong [1,b_2]$, and therefore $E \cong \langle a_1,a_2\rangle \otimes [1,b_1] \in IWq(A)_o$. We now assume $n > 2$ and use induction on n. From $\hat{a}([E]) = 1$ we conclude

$$A(\wp^{-1}(-b_1)) \circ \ldots \circ A(\wp^{-1}(-b_n)) \cong A \times A$$

On the other hand we have $-[1,b_1] \perp [1,b_2] \perp \ldots \perp [1,b_n] \cong \mathbb{H} \perp G$ with $\dim G = n-2$ and $\hat{a}([G]) = 1$ (in virtue of the above relation). Hence the induction hypothesis implies $[G] \in IWq(A)_o$. Using the equivalence $[1,b_1] \sim -G \perp [1,b_2] \perp \ldots \perp [1,b_n]$ we deduce that

$$E \sim \langle -a_1\rangle \otimes G \perp \langle a_1,a_2\rangle \otimes [1,b_2] \perp \ldots \perp \langle a_1,a_n\rangle \otimes [1,b_n]$$

that is $[E] \equiv \langle -a_1\rangle \otimes [G] (\mathrm{mod}\ IWq(A)_o)$. Since $[G] \in IWq(A)_o$, it follows that $[E] \in IWq(A)_o$. This concludes the proof of the theorem.

(2.6) Corollary. For any semi local ring A it holds that

$$Wq(A)_o / IWq(A)_o \xrightarrow{\sim} \Delta(A)$$

(2.7) Remark. Using remark (1.11), chap. II we get for a semi local ring A with $4 = o$ the following splitting exact sequence

$$o \to IWq(A)_o \to Wq(A) \xrightarrow{a} A/\wp(A) \to o$$

where we have identified $\Delta(A)$ with $A/\wp(A)$. The section $s : A/\wp(A) \to Wq(A)$ is defined by $s(\bar{b}) = [[1,-b]]$. To prove that s is a group homomorphism, we must show the relation $[1,b] \perp [1,c] \sim [1,b+c]$ for all $b,c \in A$. Now, we set $\langle e,f\rangle = [1,b]$ and $\langle g,h\rangle = [1,c]$, where $q(e) = q(g) = 1$, $q(f) = b$, $q(h) = c$ and $(e,f) = (g,h) = 1$. The element $k = e + 2f + g \in \langle e,f\rangle \perp \langle g,h\rangle$ has the value $q(k) = 0$ and it is primitive, hence $[1,b] \perp [1,c]$ is isotropic. Therefore we have $[1,b] \perp [1,c] = \langle k,h \rangle \perp G$, where $e \in G$ because $(e,k) = (e,h) = 0$. Thus $G = [1,d]$ for some $d \in A$. But by comparing the Arf-invariants, we conclude $d = b + c$, that is $[1,b] \perp [1,c] \cong \mathbb{H} \perp [1,b+c]$, proving the claim.

Let us consider again a general semi local ring A. Then we have (see $[Pf]_3$).

(2.8) Proposition. For all $[E] \in I^2Wq(A)_0$ it holds that

$$\hat{a}([E]) = 1 \quad \text{and} \quad w([E]) = 1$$

Proof. Take $[E] \in I^2Wq(A)_0$. We only need to show that $w([E]) = 1$. Since the spaces $\langle\langle a,b,c]]$ with $a,b,1-4c \in A^*$ generate additively the group $I^2Wq(A)_0$, we may assume (in virtue of the results of § 3, chap. III) that $E = \langle\langle a,b,c]]$. Now a simple computation shows our assertion.

The converse of (2.8) remains still unsolved, even in the case of a field of characteristic not 2. However for fields of characteristic 2 the converse of (2.8) is true (see [Sa]). Some progress on this question has been achieved by Elman and Lam in their papers (see the references, particularly $[E\text{-}L]_3$, $[E\text{-}L]_7$). In § 4 of this chapter we shall show using similar arguments as in $[Pf]_3$ that the converse of (2.8) is true for spaces of dimension ≤ 12.

§ 3 Some computations

Let us consider a ring homomorphism $i : A \longrightarrow B$. Then i induces ring homomorphisms $i^* : Wq(A) \longrightarrow Wq(B)$ and $i^* : W(A) \longrightarrow W(B)$. If $A \longrightarrow B$ is a Frobenius extension with trace map $s : B \longrightarrow A$, we have two group homomorphisms $s_* : Wq(B) \longrightarrow Wq(A)$ and $s_* : W(B) \longrightarrow W(A)$ (see chapter I). Then for any $x \in Wq(A)$ or $x \in W(A)$ we have the well-known Frobenius reciprocity law $s_*(i^*(x)) = s_*(1)x$, where $1 \in W(B)$ is the unit element. In this section we shall perform some explicite calculations with the transfer map s_* (see $[Sch]_1$). Let $p(X) = X^n + a_{n-1} X^{n-1} + \cdots + a_0 \in A[X]$ be a monic polynomial over A and define

(3.1) $$B = A[X]/(p(X)).$$

B is a free algebra of dimension n over A with the basis $\{1,x,\ldots,x^{n-1}\}$ subjected to the relation $x^n = -a_{n-1}x^{n-1} - \cdots - a_0$. Now we define a linear map $s : B \longrightarrow A$ by $s(1) = 1$, $s(x) = \cdots = s(x^{n-1}) = 0$. Then s is a trace map if and only if $a_0 \in A^*$. This can be seen as follows. Consider the associated bilinear map $\bar{s} : B \times B \longrightarrow A$, $\bar{s}(b,c) = s(bc)$ for all $b,c \in B$. Then we easily see that the matrix of $\bar{s}$ with respect to the basis $\{1,x,\cdots,x^{n-1}\}$ is given by

(3.2) $$\begin{pmatrix} 1 & 0 & & & 0 \\ 0 & & & \cdot & -a_0 \\ & & 0 & \cdot & \\ & \cdot & \cdot & & * \\ 0 & -a_0 & & & \end{pmatrix}$$

whose determinant is $\pm a_0^{n-1}$. Therefore $(B,\bar{s})$ is a bilinear space if and only if $a_0 \in A^*$. Hence from now on we shall assume that $a_0 \in A^*$, that is $i : A \longrightarrow B$ is a Frobenius extension with trace map s. Now let us calculate $s_*(1)$ for $1 \in W(B)$. Of course $s_*(\langle 1\rangle) = (B,\bar{s})$ over A with the matrix (3.2). From (3.2) we immediately see that $s_*(\langle 1\rangle) = \langle 1\rangle \perp \langle x,\ldots,x^{n-1}\rangle$. Let us set $B_o = \langle x,\ldots,x^{n-1}\rangle$. Then the matrix of $\bar{s}$ on B_0 with respect to the above basis is given by

$$\begin{pmatrix} 0 & & & -a_0 \\ & & \cdot & \\ & \cdot & & \\ \cdot & & & * \\ -a_0 & & & \end{pmatrix}$$

Now we consider two cases:

i) $n = 2m$, i.e. $\dim B_0 = 2m-1$. We denote the dual basis to $\{x,\ldots,x^{2m-1}\}$ by $\{y_1,\ldots,y_{2m-1}\}$, that is

$$\bar{s}(x^i,y_j) = \delta_{ij} \quad \text{for} \quad 1 \le i,j \le 2m-1$$

Since the subspace $\langle x,y_1\rangle \subset B_0$ is metabolic (and in particular non singular), we have $B_o = \langle x,y_1\rangle \perp \langle x,y_1\rangle^{\perp}$. Now we easily see that $x^2,\ldots,x^{2m-2} \in \langle x,y_1\rangle^{\perp}$, that is

$$\langle x^2,\ldots,x^{2m-2}\rangle \subseteq \langle x,y_1\rangle^{\perp} .$$

Comparing the ranks we deduce that $\langle x,y_1\rangle^{\perp} = \langle x^2,\dots,x^{2m-2}\rangle$. Let us now denote the dual basis of $\{x^2,\dots,x^{2m-2}\}$ in $\langle x,y_1\rangle^{\perp}$ by $\{z_2,\dots,z_{2m-2}\}$. Again $\langle x^2,z_2\rangle$ is metabolic and we get $\langle x,y_1\rangle^{\perp} = \langle x^2,z_2\rangle \perp \langle x^2,z_2\rangle^{\perp}$. Since $x^3,\dots,x^{2m-2} \in \langle x^2,z_2\rangle^{\perp}$, we deduce as above that $\langle x^2,z_2\rangle^{\perp} = \langle x^3,\dots,x^{2m-2}\rangle$. Iterating these arguments we see that $B_o = \mathbb{M}(U) \perp \langle x^m\rangle$, for a suitable free submodule U. Thus we have

$$s_*(\langle 1\rangle) \cong \langle 1\rangle \perp \langle x^m\rangle \perp \mathbb{M}(U)$$

But $\bar{s}(1,1) = 1$ and $\bar{s}(x^m,x^m) = s(x^{2m}) = -a_o$, that is

$$s_*(\langle 1\rangle) \cong \langle 1,-a_o\rangle \perp \mathbb{M}(U),$$

and in W(A) we have

$$s_*(1) = [\langle 1,-a_o\rangle].$$

ii) $n = 2m+1$. In this case we have $\dim B_o = 2m$ and $U = \langle x,\dots,x^m\rangle \subset B_o$ is a totally isotropic subspace. Since $\dim B_o = 2\dim U$, it follows that $U = U^{\perp}$ and $B_o = \mathbb{M}(V)$ for a suitable subspace V of B_o (see (3.8), chap. I). Therefore

$$s_*(\langle 1\rangle) \cong \langle 1\rangle \perp \mathbb{M}(V)$$

and in W(A) we get $s_*(1) = 1$.
Summing up these results we have

(3.3) Proposition. Let A be a ring and consider the Frobenius extension $B = A[X]/(p(X))$, where $p(X) = X^n + \dots + a_o$, $a_o \in A^*$. Let $s\colon B \longrightarrow A$ be the trace map given by $s(1) = 1$, $s(x) = \dots = s(x^{n-1}) = 0$. Then we have for a suitable subspace V of $(B,\bar{s})$

$$s_*(\langle 1\rangle) \cong \begin{cases} \langle 1,-a_o\rangle \perp \mathbb{M}(V) & \text{if } n = 2m \\ \langle 1\rangle \perp \mathbb{M}(V) & \text{if } n = 2m+1 \end{cases}$$

(3.4) Corollary. Let $i\colon A \longrightarrow B$ be as in (3.3) and assume n to be odd. Then the homomorphisms $i^*\colon Wq(A) \longrightarrow Wq(B)$ and $i^*\colon W(A) \longrightarrow W(B)$ are both one to one.

Proof. This follows immediately from the formula $s_*(i^*(x)) = s_*(1)x$

and the fact $s_*(1) = 1$ (see (3.3)).

(3.5) Remark. If A is semi local, then the assertion of (3.4) is true for all Frobenius extensions of A. This fact will be proved in § 6 of this chapter.

(3.6) Corollary. Let $i: A \longrightarrow B$ be as in (3.3) and assume n to be even. Then it holds that

$$[<1,-a_o>] \quad Ker(i^*) = 0$$

Proof. Take $z \in Ker(i^*)$. We use again the relation $s_*(i^*(z)) = s_*(1)z$. Then our assertion follows immediately from (3.3).

(3.7) Remark. If $s,t : B \longrightarrow A$ are two trace maps for a Frobenius extension $A \longrightarrow B$, then they are related in the following way: there exists a unit $c \in B^*$ such that $s(b) = t(cb)$ for all $b \in B$. In particular $Im(s_*) = Im(t_*)$. Now let us consider again the Frobenius extension $B = A \oplus Ax \oplus \ldots \oplus Ax^{n-1}$ with $x^n + \ldots + a_o = 0$. We define the trace map $t: B \longrightarrow A$ by $t(1) = \ldots = t(x^{n-2}) = o$, $t(x^{n-1}) = 1$. Then this map is related to the map s by

$$t(y) = -a_o^{-1} s(xy)$$

for all $y \in B$. Let us in particular consider the case $B = A(\sqrt{a}) = A[X]/(X^2 - a)$ for a semi local ring A, where $2,a \in A^*$. We know that W(B) is additively generated by the one-dimensional spaces $<\alpha>$ with $\alpha \in B^*$. Let us consider $\alpha = b + c\sqrt{a} \in B^*$ with $b,c \in A$. Since $<\alpha> \cong <\alpha\beta^2>$ for any $\beta \in B^*$, we may scale α with a suitable $\beta^2 \in B^*$ to assure that $c \in A^*$. This can be seen as follows. We set $\beta = x + y\sqrt{a}$ with $x,y \in A$. Hence $\alpha\beta^2 = b(x^2 + y^2a) + 2xyac + [c(x^2 + y^2) + 2bxy]\sqrt{a}$. Since $2 \in A^*$, we can find (using the chinese remainder theorem) $x,y \in A$ such that $c(x^2 + y^2) + 2bxy \in A^*$. Thus we can assume $c \in A^*$. Taking the trace map $t: B \longrightarrow A$, $t(1) = 0$, $t(\sqrt{a}) = 1$, we get

$$t_*(<\alpha>) \cong \begin{pmatrix} c & b \\ b & ca \end{pmatrix}$$
$$\cong <c> \otimes <1, -N(\alpha)>,$$

where $N(\alpha) = b^2 - c^2a$ is the norm of α. Thus we have shown

(3.8) $$t_*[W(B)] = \sum_{\alpha \in B^*} W(A)\ [\langle 1, -N(\alpha)\rangle]$$

Since $\{N(\alpha) \mid \alpha \in B^*\} = \underline{D}(\langle 1, -a\rangle)^* = N(\langle 1, -a\rangle)$ (see § 1, chap. IV), we get in virtue of (4.6), chap. IV,

(3.9) $$\mathrm{Ann}(\langle 1, -a\rangle) = \mathrm{Im}(s_*)$$

This relation is a special case of the more general result (5.7), proved in § 5 of this chapter.

§ 4. Quadratic separable extensions.

In this section we shall throughout assume, that A is a semi local ring. Then any quadratic separable extension $i: A \longrightarrow B$ has the form (see chap. II, § 1)

(4.1) $$B = A[X]/(X^2 - X + b),$$

where $1-4b \in A^*$. Let $\delta \in B$ be the class of X. Hence $B = A \oplus A\delta$ and $\delta^2 = \delta - b$ (in this notation i is the canonical inclusion of A into $A \oplus \delta$). Over every maximal ideal m of A there exist at most two maximal ideals M_1, M_2 of B with the property $M_i \cap A = m$. If $M_1 \neq M_2$, then $M_1 \cap M_2 = mB$ and $B/M_i \cong A/m$ for $i = 1, 2$. Therefore $B/mB \cong B/M_1 \times B/M_2 \cong A/m \times A/m$ is the trivial quadratic separable A/m-algebra. On the other hand, if $M_1 = M_2$, then $M = mB$ is the unique maximal ideal of B over m, and B/mB is a quadratic separable field extension of A/m. Denoting by r the Jacobson radical of A, we see that rB is the Jacobson radical of B.
Our next aim is to compute $\mathrm{Ker}(i^*)$, where $i^*: Wq(A) \longrightarrow Wq(B)$ is induced by $i : A \longrightarrow B$.

(4.2) Theorem. Let (E,q) be a quadratic space over A with $\dim_A E \geq 3$. If $B \otimes E$ is isotropic over B, then (E,q) contains a subspace isomorphic to $\langle c\rangle \otimes [1,b]$ for some $c \in A^*$.

Proof. Let us assume that $B \otimes E$ is isotropic over B. This means that there exist $x + \delta y$, $u + \delta v \in B \otimes E$ with $x,y,u,v \in E$ such that

$$q(x + \delta y) = q(u + \delta v) = 0$$

$$(x + \delta y,\quad u + \delta v) = 1$$

These equations are equivalent with

$$(4.3)\qquad q(x) = bq(y), \qquad\qquad q(y) = -(x,y)$$

$$(4.4)\qquad q(u) = bq(v), \qquad\qquad q(v) = -(u,v)$$

$$(4.5)\qquad (x,u) - b(y,v) = 1, \qquad\qquad (x,v) + (y,v) + (y,u) = 0$$

Our next purpose is to show that there exist $x', y' \in E$ with the property (4.3) and moreover $q(y') \in A^*$. Once this is achieved, the proof of our theorem follows immediately. Because on the subspace $\langle x', y'\rangle = Ax' \oplus A(-y')$ the form q has the matrix

$$\begin{bmatrix} bq(y') & q(y') \\ q(y') & q(y') \end{bmatrix} \cong \langle q(y')\rangle \otimes \begin{bmatrix} 1 & 1 \\ 1 & b \end{bmatrix}$$

that is $\langle x', y'\rangle \cong \langle c\rangle \otimes [1,b]$, where $c = q(y')$. In particular $E = \langle x', y'\rangle \perp \langle x', y'\rangle^{\perp}$, proving the assertion of the theorem. Thus we only have to find $x', y' \in E$ with

$$(4.6)\qquad q(x') = bq(y'), \qquad\qquad q(y') = -(x',y') \in A^*.$$

Let us first reduce the problem to the field case. We choose any $m \in \max(A)$ and $M \in \max(B)$, with $M \cap A = m$. Let $\bar{\delta}$ be the image of δ in B/M. Since $x + \delta y \in B \otimes E$ is primitive (see (4.5)), we have $\bar{x} + \bar{\delta}\bar{y} \neq 0$ in $B \otimes E/M(B\otimes E)$, where $\bar{x}, \bar{y}$ are the images of x and y in E/mE, respectively. Moreover $\bar{q}(\bar{x} + \bar{\delta}\bar{y}) = 0$. Since we are assuming (4.6) to be true over fields, we can find $\bar{x}, \bar{y}' \in E/mE$ such that $\bar{q}(\bar{x}') = \bar{b}\bar{q}(\bar{y}')$ and $\bar{q}(\bar{y}') = -(\bar{x}',\bar{y}') \neq 0$. In particular $\bar{x}' + \bar{\delta}\bar{y}' \in B \otimes E/M(B \otimes E)$ is an isotropic vector $\neq 0$. But $\dim (B \otimes E) \geq 3$, thus there exists $\bar{\sigma} \in O^+(B \otimes E/M(B \otimes E))$ with (see (4.7) below)

$$\bar{\sigma}(\bar{x} + \bar{\delta}\bar{y}) = \bar{x}' + \bar{\delta}\bar{y}'\ .$$

Doing this for all $m \in \max(A)$ and $M \in \max(B)$ wo obtain an element

$$(\bar{\sigma}) \in \prod_M O^+(B \otimes E/M(B \otimes E)) = O^+(\overline{B \otimes E})\ .$$

Now, according to (3.8), chap. III, there exists a lifting $\sigma \in O^+(B\otimes E)$ of $(\bar{\sigma})$. We define $x', y' \in E$ by

$$\sigma(x + \delta y) = x' + \delta y'$$

Then for any $m \in \max(A)$ we get $x' \longrightarrow \bar{x}'$, $y' \longrightarrow \bar{y}'$ under the reduction modulo mE. Thus, in particular, we have $q(y') \in A^*$. This proves (4.6), because $x' + \delta y'$ is isotropic.

Hence from now on we shall assume that A is a field. According to (4.3), (4.5) we have to our disposal two vectors $x,y \in E$, not both zero, such that $q(x) = bq(y)$, $q(y) = -(x,y)$. We may also assume that $q(y) = 0$, and therefore $q(x) = (x,y) = 0$. Since $x \neq 0$ or $y \neq 0$, we get that E is isotropic. Therefore, there exist vectors $e,f \in E$ such that $q(e) = q(f) = 0$ and $(e,f) = 1$. We set $E = \langle e,f\rangle \perp F$, where $\dim F \geq 1$. Now let us consider the following two cases:

i) $ch(A) \neq 2$ Taking $z \in F$ with $q(z) \neq 0$ we define

$$x' = e + (b - \tfrac{1}{4})q(z)f - \tfrac{1}{2} z$$

$$y' = z$$

ii) $ch(A) = 2$. Hence $\dim F \geq 2$, and we can find $z,t \in F$ with $q(z) = a \neq 0$, $q(t) = d$ and $(z,t) = 1$. Then we define

$$x' = e + a(b - ad)f + at$$

$$y' = z$$

In both cases, the pair $\{x', y'\}$ satisfies(4.6). This concludes the proof of the theorem.

During the proof of (4.2) we have used the following result.

(4.7) <u>Lemma</u>. Let (E,q) be a quadratic space over the field A. Assume $\dim_A E \geq 3$. If $x,y \in E$ are two non zero elements with $q(x) = q(y)$, then there exists $\sigma \in O^+(E)$, such that $\sigma(x) = y$.

<u>Proof</u>. According to Witt's theorem (see (4.1), chap. III), the isometry $\langle x\rangle \longrightarrow E$, defined by $x \longrightarrow y$, has an extension $\sigma' \in O(E)$. Since $y \neq 0$, we can easily find a two-dimensional non singular sub-

space E_o of E, which contains y. Hence $E = E_o \perp E_1$ and $\dim_A E_1 \geq 1$. If $\sigma' \in O^+(E)$, we simply take $\sigma = \sigma'$. If $\sigma' \notin O^+(E)$, we pick $z \in E$, with $q(z) \neq 0$, and define $\sigma = \sigma_z\sigma'$. Then $\sigma \in O^+(E)$ and $\sigma(x) = y$.

We don't know whether this result is true or not true in the semi local case.

Using (4.2), we deduce by induction on dim E.

(4.8) <u>Corollary</u>. Let (E,q) be a quadratic space over A. If $B \otimes E$ is hyperbolic over B (see (4.1)), then there exists a bilinear space $\varphi = \langle c_1,\ldots,c_n\rangle$ (with $c_i \in A^*$) and $a,d \in A$ with $d, 1-4a \in A^*$, such that

$$E \cong \varphi \otimes [1,b] \perp \langle d\rangle \otimes [1,a],$$

where $$B \otimes [1,a] \cong [0,0] \text{ over } B.$$

<u>Proof</u>. In case $\dim_A E = 2$ there is nothing to prove. Thus let us assume that $\dim_A E \geq 3$. Since $B \otimes E$ is hyperbolic, and in particular isotropic, we conclude from (4.2) that

$$E \cong \langle c_1\rangle \otimes [1,b] \perp E ,$$

with $E_1 \neq 0$. Hence $B \otimes E \cong \mathbb{H} \perp B \otimes E_1$, because $B \otimes [1,b] \cong \mathbb{H}$. Now using the cancelation theorem (4.3), chap. III, we get that $B \otimes E_1$ is hyperbolic. Thus we can apply induction to E_1.

According to (4.8), we are led to characterise all quadratic spaces [1,a] over A with $[1,a] \otimes B \cong \mathbb{H}$. This is possible if A is assumed to be <u>connected</u>, that is the only idempotents of A are 0 and 1.

(4.9) <u>Lemma</u>. Let A be a connected semi local ring. If [1,a] is a quadratic space over A with $B \otimes [1,a] \cong [0,0]$, then either $[1,a] \cong [0,0]$ or $[1,a] \cong [1,b]$.

<u>Proof</u>. Let us assume $B \otimes [1,a] \cong [0,0]$. Then there exist $\alpha,\beta \in B$ with $\alpha^2 + \alpha\beta + a\beta^2 = 0$ and (α,β) primitive. It follows that $\beta \in B^*$, because if $\beta \in M$ for some $M \in \max(B)$, then from the above equation we get $\alpha \in M$ too, which is a contradiction. Therefore, dividing the above equation by β we may assume that $\beta = 1$, that is $\alpha^2 + \alpha + a = 0$. Now we consicer $A(\wp^{-1}(-a)) = A \oplus Az$ with $z^2 = z - a$. The above relation defines a homomorphism of algebras

$$f : A(\wp^{-1}(-a)) \longrightarrow B$$

by $f(z) = -\alpha$. But both algebras are separable over A, thus it follows that the kernel of f is generated by an idempotent, i.e. $\mathrm{Ker}(f) = A(\wp^{-1}(-a)) \cdot e$, where $e^2 = e$, $e \neq 1$ (see (2.6), chap. III in [I-DeM]). Assuming that $A(\wp^{-1}(-a))$ is not connected, i.e. has non trivial idempotents, we easily see that $A(\wp^{-1}(-a)) \cong A \times A$, because A is connected. In this case we have $[1,a] \cong [0,0]$. On the other hand, if $A(\wp^{-1}(-a))$ is connected, we have $\mathrm{Ker}(f) = 0$, and comparing the ranks of B and $f[A(\wp^{-1}(-a))]$, we deduce that f is an isomorphism. In this case we have $[1,a] \cong [1,b]$. This concludes the proof of the lemma.

Clearly, the above result is not true in general for not connected semi local rings.

(4.10) Corollary. Let A be a connected semi local ring. If (E,q) is a quadratic space over A with $B \otimes E$ hyperbolic, then there is a bilinear space $\varphi = \langle c_1,\dots,c_n\rangle$, $c_i \in A^*$, such that either

$$E \cong \varphi \otimes [1,b] \quad \text{or} \quad E \cong \varphi \otimes [1,b] \perp [0,0]$$

(4.11) Corollary. Let A be a semi local ring. Then for the quadratic separable extension $i : A \longrightarrow B$ we have

$$\mathrm{Ker}(i^*) = W(A)[1,b]$$

Proof. If A is connected, the corollary follows immediately from (4.10). If A is non connected, then $A = A_1 \times \dots \times A_s$ is the decomposition of A in connected components. Then $Wq(A) = Wq(A_1) \times \dots \times Wq(A_s)$. All things over A decompose according to the above decomposition of A, so that for example $B = B_1 \times \dots \times B_s$, where B_i is a quadratic separable A_i-algebra. Thus the induced homomorphism $i^*: Wq(A) \longrightarrow Wq(B)$ is the product of the homomorphisms $i_k: Wq(A_k) \longrightarrow Wq(B_k)$ and therefore $\mathrm{Ker}(i^*) = \prod_k \mathrm{Ker}(i^*_k)$. Now our assertion follows directly from (4.10).

Now the homomorphism $i^*: Wq(A) \longrightarrow Wq(B)$ induces for any integer $n \geq 0$ homomorphisms

$$i^*_n : I^n_A Wq(A)_0 \longrightarrow I^n_B Wq(B)_0$$

Then the following result is a refinement of (4.11).

(4.12) Proposition. For $0 \leq n \leq 2$ we have

$$\mathrm{Ker}(i_n^*) = I_A^n\,[1,b]$$

Proof. We may assume that A is connected, and since the case $B = A \times A$ is obvious, we shall assume B to be connected, too. Since the case $n = 0$ was considered in (4.11), we may suppose that $n \geq 1$. Let us now assume that $[q] \in I_A\, Wq(A)_0$ and $i^*([q]) = 0$, i.e. $B \otimes q \sim 0$. Then there is a bilinear space φ over A, $\varphi = \langle a_1,\dots,a_m\rangle$, such that $q \sim \varphi \otimes [1,b]$. Since $a([q]) = 1$ and $a(\varphi \otimes [1,b]) = [A(\wp^{-1}(-b))]^m$, it follows that m must be even. In particular $\varphi \in I_A$, proving the case $n = 1$. Let us take now $[q] \in I_A^2\, Wq(A)_0$, such that $q \otimes B \sim 0$, that is $q \sim \varphi \otimes [1,b]$ for a suitable $\varphi = \langle a_1,\dots,a_m\rangle$. By the above argument we have $m = 2s$. Hence we can write $\varphi \sim \langle a_1\rangle \otimes \langle\langle a_1a_2, a_1a_3\rangle\rangle \perp \varphi_1$ with $\dim \varphi_1 = 2s - 2$. Therefore

$$q \sim \langle a_1\rangle \otimes \langle\langle a_1a_2,\ a_1a_3,\ b]] \perp \varphi_1 \otimes [1,b]$$

and in particular $\varphi_1 \otimes [1,b] \in I_A^2\, Wq(A)_0$. Now we use the remark (4.14), (1), below to conclude that $\varphi_1 \otimes [1,b] \sim \varphi_2 \otimes [1,b]$ with a suitable $\varphi_2 \in I_A^2$. Hence

$$q \sim (\langle a_1\rangle \otimes \langle\langle a_1a_2,\ a_1a_3\rangle\rangle \perp \varphi_2) \otimes [1,b],$$

that is $q \sim \varphi_3 \otimes [1,b]$ with $\varphi_3 \in I_A^2$. This proves the proposition.

(4.13) Remark. 1) Let φ be a bilinear space, such that $\varphi \otimes [1,b] \in I_A^2\, Wq(A)_0$. Then there exists $\varphi_1 \in I_A^2$, such that $\varphi \otimes [1,b] \sim \varphi_1 \otimes [1,b]$. Of course φ must be even dimensional, and we may assume $\varphi = \langle a_1,\dots,a_{2s}\rangle$. If $s = 1$, then our assertion is clear, as we can see by computing invariants. Thus let us assume $s > 1$. We use again the decomposition $\varphi \sim \langle a_1\rangle \otimes \langle\langle a_1a_2, a_1a_3\rangle\rangle \perp \varphi_0$, where $\dim \varphi_0 = 2s - 2$, to conclude that $\varphi_0 \otimes [1,b] \in I_A^2[1,b]$. Our assertion follows now by induction.

2) The assertion of (4.2) allows the following interpretation. For any quadratic space (E,q) over A is the anisotropic part of $(B \otimes E)$ defined over A, i.e. $B \otimes E \cong r \times \mathbb{H} \perp B \otimes E_0$, $r \geq 0$, for some quadratic space (E_0,q_0) over A, such that $B \otimes E_0$ is anisotropic over B. In par-

ticular, it follows that for any $[q] \in \mathrm{Im}(i^*)$ with $\dim q > 0$, there exists $d \in \underline{D}(B \otimes q) \cap A^*$ (see $[E\text{-}L]_8$). Using this remark one can prove the following

(4.14) Proposition. Let q be a quadratic Pfister space over B such that $q \cong q_0 \otimes B$ for some quadratic space q_0 over A. Then there is a quadratic Pfister space q_1 over A such that $q \cong q_1 \otimes B$.

Proof. Writing $q_o = \langle c\rangle \otimes [1,a] \perp \ldots$, we get $c \in \underline{D}(q \otimes B)^*$, and since q is round we may replace q_0 by $\langle c\rangle \otimes q_0$. Hence we may assume that $c = 1$. In particular q contains the quadratic Pfister space $[1,a]$, where $a \in A$. Let now q_1 be a quadratic Pfister space over A of maximal dimension, such that $q_1 \otimes B \subseteq q$. Using (3.4), chap. IV, we conclude that

$$q \cong \tau \otimes (q_1 \otimes B)$$

for a suitable bilinear Pfister space τ over B. Let us assume that $\dim \tau > 1$, and write $\tau = \langle 1\rangle \perp \tau'$ with $\dim \tau' \geq 1$. Hence $q \cong q_1 \otimes B \perp \tau' \otimes q_1 \otimes B$, and therefore $[\tau' \otimes q_1 \otimes B] \in \mathrm{Im}(i^*)$. Using the remark above we see that $\tau' \otimes q_1 \otimes B$ represents an element $d \in A^*$ over B. Hence (3.3), chap. IV, implies

$$q \cong \tau \otimes (q_1 \otimes B) \cong \tau_1 \otimes \langle 1,d\rangle \otimes q_1 \otimes B$$

for a suitable bilinear Pfister space τ_1. In particular $\langle 1,d\rangle \otimes q_1 \otimes B$ is a subspace of q, which is a contradiction. Hence $\tau = \langle 1\rangle$ and $q \cong q_1 \otimes B$. This proves our assertion.

We now prove a weaker, but rather more general, version of (4.13). Let us assume that q is a quadratic space over B, which is defined over A. Let us assume that $q \cong \varphi \otimes (q_0 \otimes B)$, where q_0 is a Pfister quadratic space over A and $\varphi = \langle c_1,\ldots,c_n\rangle$ is a bilinear space over B. Then in virtue of the remark (4.14), (2), we can find $d \in \underline{D}(q) \cap A^*$. Therefore we have $d = c_1q_0(x_1) + \ldots + c_nq_0(x_n)$, $x_i \in q_0 \otimes B$. We can assume $q_0(x_i) \in B^*$ for all i, provided $|B/M| \geq 3$ for all $M \in \max(B)$ or that n is even (see (5.2), chap. III). Let us assume one of this conditions. Since q_0 is round (also over B), we get

$$\begin{aligned} q &= \langle c_1,\ldots,c_n\rangle \otimes q_o \cong \langle c_1q_o(x_1),\ldots,c_nq_o(x_n)\rangle \otimes q_o \\ &\cong \langle d,d_2,\ldots,d_n\rangle \otimes q_o \end{aligned}$$

for some $d_2,\dots,d_n \in B^*$. In particular $\langle d_2,\dots,d_n\rangle \otimes q_0$ is defined over A, thus we can use induction to prove the following.

(4.15) Proposition. Let q be a quadratic space over B, which is defined over A and such that $q \cong \varphi \otimes q_0 \otimes B$ for some quadratic Pfister space q_0 over A, and $\varphi = \langle c_1,\dots,c_n\rangle$, $c_i \in B^*$. Assume either $|B/M| \geq 3$ for all $M \in \max(B)$ or $n \equiv 0(2)$. Then there exists a bilinear space $\psi = \langle a_1,\dots,a_n\rangle$ with $a_i \in A^*$, such that $q \cong (\psi \otimes q_0) \otimes B$.

We make an application of these results. Let us consider $a,b \in A$ with $1-4a$, $1-4b \in A^*$. We define the biquadratic extension $C = A(\wp^{-1}(-a), \wp^{-1}(-b)) = A(\wp^{-1}(-a))(\wp^{-1}(-b))$ of A. Let $j : A \longrightarrow C$ be the natural inclusion and $j^*: W_q(A) \longrightarrow Wq(C)$ the induced homomorphism between the corresponding Witt rings. Then we have (see $[E-L]_8$).

(4.16) Corollary. Lat A be a semi local ring with $|A/m| \geq 3$ for all $m \in \max(A)$. Then

$$\mathrm{Ker}(j^*) = W(A)[1,a] + W(A)[1,b]$$

Proof. Consider $[q] \in \mathrm{Ker}(j^*)$. According to (4.11) we may assume that $q \otimes A(\wp^{-1}(-a))$ is anisotropic, but $q \otimes C \sim 0$. Again in virtue of (4.11) we have

$$q \otimes A(\wp^{-1}(-a)) \sim \varphi \otimes [1,b]$$

for a suitable bilinear space φ over $A(\wp^{-1}(-a))$. Since $\varphi \otimes [1,b]$ is defined over A, we can use (4.15) to assume that φ is defined over A, too. Hence $[q \perp - \varphi \otimes [1,b]] \in \mathrm{Ker}(i^*)$, where $i : A \longrightarrow A(\wp^{-1}(-a))$ is the natural inclusion. Now we use again (4.11) to finish the proof of our assertion.

We now want to generalize for semi local rings the well-known result of Pfister, that for any quadratic space q with $\dim q \leq 12$, $\hat{a}(q) = 1$, $w(q) = 1$, necessarily follows $[q] \in I^2Wq(A)_0$ (see $[Pf]_3$). In the sequel A denotes a connected semi local ring. Consider a class $[q] \in Wq(A)$ such that $\dim q \leq 12$, $\dim q \equiv 0 \pmod 2$, $a(q) = 1$ and $w(q) = 1$. We assert that $[q] \in I^2Wq(A)_0$. Following $[Pf]_3$, we distinguish several cases:

1) dim q = 2. Hence $q = \langle a\rangle \otimes [1,b]$ and $a(q) = [A(\wp^{-1}(-b))] = 1$, that is $[1,b] \cong \mathbb{H}$.

2) <u>$\dim q = 4$</u>. Hence $q = \langle a\rangle \otimes [1,b] \perp \langle c\rangle \otimes [1,d]$. Since $a(q) = 1$, we deduce that $[1,b] \cong [1,d]$, i.e. $q \cong \langle a\rangle \otimes \langle\langle ac,b]]$. Since $w(q) = [(-ac,-b]] = 1$, it follows that $q \sim 0$. Hence $[q] = 0$.

3) <u>$\dim q = 6$</u>. We write $q = \langle a\rangle \otimes [1,b] \perp q_0$, where $\dim q_0 = 4$. Set $B = A(\wp^{-1}(-b))$. Then $[q \otimes B] = [q_0 \otimes B]$ and using the step (2) we get $[q_0 \otimes B] = 0$ in $Wq(B)$. From (4.10) we deduce $q \cong \langle a,c'\rangle \otimes [1,b] \perp \mathbb{H}$ or $q \cong \langle a,c',d\rangle \otimes [1,b]$ for suitable $c',d \in A^*$. In the first case, we get again from step (2) that $q \sim 0$. In the second case we compute $a(q) = [A(\wp^{-1}(-b))]^3 = 1$, i.e. $[1,b] \cong \mathbb{H}$. Hence $q \sim 0$. We formulate this result separately for later references.

(4.17) <u>Proposition</u>. Every quadratic space over A of dimension 6, whose Arf- and Witt invariants are trivial, is hyperbolic.

Before we treat the rest cases, we now consider a little more general situation as in (4.17). Let q be a quadratic space over A of dimension 6, such that $a(q) = 1$ and $w(q) = [(c,d]] \in Br(A)$ for a suitable quaternion algebra $(c,d]$ over A. We set $B = A(\wp^{-1}(-d))$. Since $(c,d]$ splits over B, we can use (4.17) to conclude that $[q \otimes B] = 0$ in $Wq(B)$. Using (4.10) again we get $q \cong \langle a,b\rangle \otimes [1,d] \perp \mathbb{H}$ or $q \cong \langle a,b,e\rangle \otimes [1,d]$ with suitable $a,b,e \in A^*$. The latter case implies $a(q) = [A(\wp^{-1}(-d))]^3 = 1$, that is $[1,d] \cong \mathbb{H}$ and consequently $q \cong 3 \times \mathbb{H}$. Thus we have proved

(4.18) <u>Proposition</u>. Every quadratic space over A with $\dim q = 6$, $a(q) = 1$ and $w(q) = [(c,d]] \in Br(A)$ is isotropic.

4) <u>$\dim q = 8$</u>. We set $q = \langle a\rangle \otimes [1,b] \perp q_0$, where $\dim q_0 = 6$ and consider $B = A(\wp^{-1}(-b))$. Then $[q \otimes B] = [q_0 \otimes B]$ and we can use (4.17) to conclude that $[q_0 \otimes B] = 0$. Using (4.10) we get either $q \cong \langle a_1,a_2,a_3\rangle \otimes [1,b] \perp \mathbb{H}$ or $q \cong \langle a_1,a_2,a_3,a_4\rangle \otimes [1,b]$. The former case implies $a(q) = [A(\wp^{-1}(-b))]^3 = 1$, and therefore $[1,b] \cong \mathbb{H}$, i.e. $q \cong 4 \times \mathbb{H}$. In the latter case we define $q_1 = \langle a_1,a_2\rangle \otimes [1,b] = \langle a_1\rangle \otimes \langle\langle c_1,b]]$ and $q_2 = \langle a_3,a_4\rangle \otimes [1,b] = \langle a_3\rangle \otimes \langle\langle c_2,b]]$. Hence $q \cong q_1 \perp q_2$. It follows $a(q_1) = a(q_2)$ and $w(q_1) = w(q_2)$. This implies $q_2 \cong \langle d\rangle \otimes q_1$ with some $d \in A^*$, and hence $q \cong \langle 1,d\rangle \otimes q_1 \cong \langle a_1\rangle \otimes \langle\langle d,c_1,b]]$. Hence in any case $[q] \in I^2Wq(A)_0$.

5) <u>$\dim q = 10$</u>. We set $q = \langle a_1\rangle \otimes [1,b_1] \perp \langle a_2\rangle \otimes [1,b_2] \perp q_0$ with $\dim q_0 = 6$. Considering the isomorphism $[1,b_1] \perp -[1,b_2] \cong \mathbb{H} \perp \langle c\rangle \otimes [1,b_3]$ for suitable $c,b \in A$, we define $B = A(\wp^{-1}(-b_3))$.

Then $[1,b_1] \otimes B \cong [1,b_2] \otimes B$ and therefore $q \otimes B \cong \langle a_1,a_2\rangle \otimes [1,b_1] \otimes B \perp q_0 \otimes B$. Since $a(q) = 1$ and $w(q) = 1$, a straightforward calculation shows that $a(q_0 \otimes B) = 1$ and $w(q_0 \otimes B) = [(-a_1a_2, -b_1]] \in Br(A)$. Then (4.18) implies that $q_0 \otimes B$ is isotropic. We now use (4.2) to deduce that $q_0 \cong \langle a_3\rangle \otimes [1,b_3] \perp q_1$ and hence $q \cong \langle a_1\rangle \otimes [1,b_1] \perp \langle a_2\rangle \otimes [1,b_2] \perp \langle a_3\rangle \otimes [1,b_3] \perp q_1$. This implies $a(q_1) = 1$, and hence $q_1 \cong \langle a_4\rangle \otimes \langle\langle c,d]]$ with suitable $a_4,c,d \in A$. We define $q' = \perp_{i=1}^{3} \langle a_i\rangle \otimes [1,b_i]$, i.e, $q \cong q' \perp q_1$. Computing invariants, we get $a(q') = 1$, $w(q') = w(q_1) = [(-c,-d]]$, that is q' is isotropic (see (4.18)). In particular q itself is isotropic, so that we may use the step (4) to show that $[q] \in I^2Wq(A)_o$.

6) <u>dim q = 12</u>. Take $q = \langle a_1\rangle \otimes [1,b_1] \perp q_0$ and define $B = A(\wp^{-1}(-b_1))$. Then $a(q_0 \otimes B) = 1$ and $w(q_0 \otimes B) = 1$. Since we have shown in step (5) that any 10-dimensional quadratic form with trivial invariants is isotropic, we see that $q_0 \otimes B$ is isotropic. Now use (4.2) to conclude that $q_0 \cong \langle a_2\rangle \otimes [1,b_1] \perp q_1$, where $\dim q_1 = 8$. Hence $q \cong \langle a_1,a_2\rangle \otimes [1,b_1] \perp q_1$. We easily compute $a(q_1) = 1$, $w(q_1) = [(-a_1a_2,-b_1]]$. We set $q_1 = \langle a_3\rangle \otimes [1,b_2] \perp q_2$ and define $B' = A(\wp^{-1}(-b_2))$. Then $a(q_2 \otimes B') = 1$, $w(q_2 \otimes B') = [(-a_1a_2,-b_1]]$. Using (4.18) we conclude that $q_2 \otimes B'$ is isotropic, since $\dim q_2 = 6$. Then $q_2 \cong \langle a_4\rangle \otimes [1,b_2] \perp q_3$ (see (4.2)). Therefore

$$q \cong \langle a_1,a_2\rangle \otimes [1,b_1] \perp \langle a_3,a_4\rangle \otimes [1,b_2] \perp q_3$$

Since $a(q_3) = 1$, we can write $q_3 = \langle a_5,a_6\rangle \otimes [1,b_3]$. Then we define $q' = \langle a_1,a_2\rangle \otimes [1,b_1]$, $q'' = \langle a_3,a_4\rangle \otimes [1,b_2]$, and we get $a(q') = a(q'') = a(q_3) = 1$, and $w(q')w(q'')w(q_3) = 1$. Putting $\bar{q} = \langle a_3\rangle \otimes q' \perp \langle -a_1\rangle \otimes q'' \perp q_3$, we get $a(\bar{q}) = 1$, $w(\bar{q}) = 1$ and since $\bar{q}$ is isotropic, we can use step (5) to conclude that $[\bar{q}] \in I^2Wq(A)_o$. On the other hand we have

$$q \sim \langle 1,-a_3\rangle \otimes q' \perp \langle 1,a_1\rangle \otimes q'' \perp \bar{q}$$

showing that $[q] \in I^2Wq(A)_o$.

We now use (4.18) to prove the following

(4.19) <u>Proposition</u>. Let $q_i = \langle\langle a_i,b_i]]$ $(i = 1, 2, 3)$ be three quaternion forms over A, such that $w(q_1)w(q_2)w(q_3) = 1$. Then there is $d \in A^*$, such that

$$q_1 \perp -q_2 \perp \langle d\rangle \otimes q_3 \cong 6 \times \mathbb{H}$$

Proof. Since $q = q_1 \perp -q_2$ is isotropic, we have $q \cong \mathbb{H} \perp q_o$, where $\dim q_o = 6$. A simple calculation shows that $a(q_o) = 1$ and $w(q_o) = w(q_1)w(q_2) = w(q_3)$, that is $w(q_o)$ us represented by a quaternion algebra. Therefore q_o is isotropic (see (4.18)), and we obtain $q \cong 2 \times \mathbb{H} \perp q'$, where $\dim q' = 4$ and $a(q') = 1$. Hence $q' \cong \langle a\rangle \otimes \langle\langle c,b]]$ for some $a,c,b \in A$, and since $w(q') = w(q) = w(q_3)$, we can find $d \in A^*$ with $q' \cong \langle -d\rangle \otimes q_3$. Now the proposition follows immediately.

Using this result and theorem (4.21) below, we get (see $[Pf]_3$).

(4.20) Corollary. Let q_i $(i = 1, 2, 3)$ be three quaternion forms over A with the property $w(q_1)w(q_2)w(q_3) = 1$. Then there exists a quadratic separable extension B of A, which splits q_1, q_2 and q_3.

Our purpose is now to prove the following.

(4.21) Theorem. Let $Q_i = (a_i, b_i]$ $(i = 1, 2)$ be two quaternion algebras over A. If $Q_1 \otimes Q_2 \sim Q_3 = (a_3, b_3]$, a third quaternion algebra, then Q_1 and Q_2 contain a common quadratic separable extension of A.

For the proof of (4.21) we need the following lemma.

(4.22) Lemma. Let $q_i = \langle\langle a_i, b_i]]$ $(i = 1, 2)$ be two quaternion quadratic forms over A, such that $q_1 \perp -q_2$ contains at least 2 hyperbolic planes. Then there exists $d \in A$ with $1-4d \in A^*$ and $c_1, c_2 \in A^*$, such that $q_i \cong \langle\langle c_i, d]]$ for $i = 1, 2$.

Proof. Let us assume $q_1 \perp -q_2 = 2 \times \mathbb{H} \perp q'$. Using the isomorphism

$$[1,b_1] \perp -[1,b_2] \cong \mathbb{H} \perp \langle 1-4b_2\rangle \otimes [1, b_1+b_2 - 4b_1b_2]$$

and the cancellation theorem (4.3), chap. III, we deduce that the quadratic space

$$\langle 1-4b_2\rangle \otimes [1,b_1+b_2-4b_1b_2] \perp \langle a_1\rangle \otimes [1,b_1] \perp - \langle a_2\rangle \otimes [1,b_2]$$

is isotropic. Then there exist $\alpha, \beta \in A$, $x_i \in \langle a_i\rangle \otimes [1,b_i]$ $(i = 1,2)$, such that $(\alpha, \beta, x_1, x_2)$ is primitive and

$$(4.23) \qquad (1-4b_2)[\alpha^2 + \alpha\beta + (b_1+b_2-4b_1b_2)\beta^2] + q_1(x_1) - q_2(x_2) = 0$$

Rearranging terms, we get

$$(1-4b_2)^2\beta^2 b_1 + q_1(x_1) = (\alpha + 2\beta b_2)^2 - (\alpha + 2\beta b_2)(\beta + 2\alpha) + (\beta + 2\alpha)^2 b_2 + q_2(x_2).$$

Let us set $\langle e_i, f_i\rangle = [1, b_i] \subset q_i$ with $q_i(e_i) = 1$, $q_i(f_i) = b_i$, $(e_i, f_i) = 1$ for $i = 1, 2$. Then we define

$$z_1 = (4b_2 - 1)\beta f_1 + x_1 \in q_1$$

$$z_2 = (\alpha + 2\beta b_2)e_2 - (\beta + 2\alpha)f_2 + x_2 \in q_2$$

Now we easily see that $q_1(z_1) = q_2(z_2)$ and $(e_1, z_1) = (e_2, z_2) = (4b_2 - 1)\beta$. This implies that the quadratic modules $\langle e_i, z_i\rangle \subset q_i$ $(i = 1, 2)$ are isomorphic, because they have with respect to the basis $\{e_i, z_i\}$ the value matrix

$$(4.24) \qquad \begin{bmatrix} 1 & (4b_2 - 1)\beta \\ (4b_2 - 1)\beta & \gamma \end{bmatrix}$$

with $\gamma = q(z_1) = q(z_2)$. Since the representation (4.23) of zero is primitive, we can apply the method used in the proof of (5.2), chap. III, to find a representation as in (4.23) with the property

$$(4.25) \qquad (4b_2 - 1)^2\beta^2 - 4\gamma \in A^*$$

Let us assume this fact without proof for a while. (4.25) means that the isomorphic modules $\langle e_1, z_1\rangle$, $\langle e_2, z_2\rangle$ are non singular, and therefore we get the splittings $(i = 1, 2)$

$$q_i \cong \langle e_i, z_i\rangle \perp p_i$$

where $\dim p_i = 2$. Since the quadratic form (4.24) represents 1 and is now non singular, it has the form $[1,d]$ for a suitable $d \in A$ with $1-4d \in A^*$. Hence $q_i \cong [1,d] \perp p_i$. Using $a(q_i) = 1$ we get $a(p_i) = [A(\wp^{-1}(-d))]$, that is $p_i \cong \langle c_i\rangle \otimes [1,d]$ for some $c_i \in A^*$. This proves $q_i \cong \langle\langle c_i, d]]$ for $i = 1, 2$. It remains to prove (4.25), to complete the proof of the lemma. Using the theorem (3.8), chap. III, we can easily reduce the proof of (4.25) to the case that A is a field. Thus we assume for the rest of the proof that A is a field. The condition (4.25) is now equivalent with

(4.26) $$(4b_1 - 1)(4b_2 - 1)^2\beta^2 + 4q_1(x_1) \neq 0$$

Suppose the contrary. Set $<e,f> = (1-4b_2) \otimes [1,b_1+b_2-4b_1b_2]$ and $x = \alpha e + \beta f + x_1 + x_2$. We shall find $\sigma \in O^+(q)$, such that $\sigma(x) = \gamma e + \delta f + y_1 + y_2$ and $(4b_1 - 1)(4b_2 - 1)^2\delta^2 + 4q_1(y_1) \neq 0$ (here $q = <e,f> \perp <a_1> \otimes [1,b_1] \perp -<a_2> \otimes [1,b_2]$). Of course it is enough to find $\sigma \in O(q)$ with this property, since we can eventually change σ through an automorphism of $-<a_2> \otimes [1,b_2]$. Now we distinguish two cases:

i) char (A) $\neq$ 2. Using (5.2), chap. III, we may assume that $\alpha e + \beta f$, x_1,x_2 are anisotropic. If $|A| \geq 5$, we choose $y_2 \in <a_2> \otimes [1,b_2]$ with $(x_2,y_2) = 0$ and $q_2(y_2) \neq 0$. Since $|A| \geq 5$, we can find $\lambda \in A^*$ with $q_1(x_1) - \lambda^2 q_2(y_2) \neq 0$. Now we define $\sigma = \sigma_{x_1+\lambda y_2}$. It is easy to see, that this σ does the job. Assume now $A = \mathbb{F}_3$. Since we are assuming (4.26) to be false, we obtain $\beta \neq 0$ and $q(x_1) = 1 - b_1$. Now we have two possibilities for b_1, namely $b_1 = 0$ or $b_1 = -1$. If $b_1 = 0$, we choose $y_1 \in <a_1> \otimes [1,b_1]$ with $q(y_1) = 0$ and $(x_1,y_1) = 1$, and $y_2 \in <a_2> \otimes [1,b_2]$ with $q(y_2) \neq 0$ and $(x_2,y_2) = 0$. Then $\sigma = \sigma_{y_1+y_2}$ has the required property. If $b_1 = -1$, we look for $y_1 \in <a_1> \otimes [1,b_1]$ with $q_1(y_1) \neq 0$ and $(x_1,y_1) = 0$. Since $[1,-1]$ is anisotropic over $\mathbb{F}_3$, we must have $q_1(y_1) = -1$. In this case $\sigma = \sigma_{y_1+x_2}$ has the required property.

ii) char (A) = 2. Since (4.26) is false, we get $\beta = 0$. In virtue of (5.1), chap. III, we can assume that $\alpha, x_1, x_2 \neq 0$. If $|A| \geq 4$ we choose $\lambda \in A^*$, such that $q(e + \lambda f) \neq 0$ and we define $\sigma = \sigma_{e+\lambda f}$. If $A = \mathbb{F}_2$, we easily see that either $q_1(x_1) = 1$ or $q_2(x_2) = 1$. Let us assume without restriction that $q_1(x_1) = 1$. Then we define $\sigma = \sigma_{f+x_1}$ if $q(f) = 0$ and $\sigma = \sigma_{e+f}$ if $q(f) = 1$. Now it is easy to see that σ, as defined above, has the required property. This completes the proof of the lemma.

<u>Proof of (4.21)</u>. Let us assume $Q_1 \otimes Q_2 \sim Q_3$. The norm form of Q_i is $q_i = <<-a_i,-b_i]]$ and the above relation means $w(q_1)w(q_2)w(q_3) = 1$ in Br(A). Now using (4.19) we conclude that $q_1 \perp -q_2$ contains two hyperbolic planes, and therefore in virtue of (4.22) we can find $c_1,c_2 \in A^*$ and $d \in A$ with $1 + 4d \in A^*$, such that $q_i \cong$

$\langle\langle -c_i, -d]]$ for $i = 1, 2$. Hence $w(q_i) = [(c_i, d]] = [Q_i]$, that is $Q_i \sim (c_i, d]$ for $i = 1, 2$. Now we use the cancelation theorem for Azumaya algebras over semi local rings (see [Kn]) to deduce that $Q_i \cong (c_i, d]$, proving that the quadratic separable extension $A(\wp^{-1}(d))$ is contained in Q_i and Q_2. This proves (4.21).

(4.26) Remark. Under the hypothesis of theorem (4.21) one would expect to find $c \in A^*$, $d_1, d_2 \in A$ with $1 + 4d_1$, $1 + 4d_2 \in A^*$, such that $Q_1 \cong (c, d_1]$ and $Q_2 \cong (c, d_2]$. Now we give a counterexample to this assertion. Let us consider a field k_o of characteristic 2, such that there is an anisotropic 2-fold quadratic Pfister form $q = \langle\langle a,b]]$ over k_o. Define $k = k_o(X)$ and $Q_1 = (X,b]$, $Q_2 = (aX,b]$. Hence the assumption in (4.21) is fulfilled for Q_1 and Q_2 over k. Let us now assume that there exist $f \in k^*$, $g_1, g_2 \in k$, such that $Q_1 \cong (f, g_1]$ and $Q_2 = (f, g_2]$. Taking the norms of these algebras we get $\langle\langle X,b]] \cong \langle\langle f, g_1]]$ and $\langle\langle aX,b]] \cong \langle\langle f, g_2]]$, and consequently $\langle\langle X,b]] \perp \langle\langle aX,b]] \cong \langle\langle f,g_1]] \perp \langle\langle f,g_2]]$. Since $2 = 0$, we get

$$2 \times \mathbb{H} \perp \langle X\rangle \otimes \langle\langle a,b]] \cong 2 \times \mathbb{H} \perp \langle\langle f, g_1 + g_2]] ,$$

and cancelling $\langle X\rangle \otimes \langle\langle a,b]] \cong \langle\langle f, g_1 + g_2]]$. In particular, $\langle\langle a,b]]$ must represent X over $k = k_o(X)$, and therefore over $k_o[X]$ (the well-known theorem of Cassels-Pfister is also true in characteristic 2). Now a simple computation shows, that q must be isotropic, contradiction.

§ 5. An exact sequence of Witt groups

The aim of this section is to prove the exactness of a sequence of Witt groups, which is associated to a quadratic separable extension of a semi local ring A (compare [L], [Ma] for special cases). In the sequel let A be a semi local ring and let $B = A(\wp^{-1}(t)) = A \oplus A\delta$ be a quadratic separable extension of A, where $\delta^2 = \delta + t$ and $1+4t \in A^*$. In § 4 we have shown that the sequence

$$(5.1) \qquad o \longrightarrow W(A)[1,-t] \longrightarrow Wq(A) \xrightarrow{i^*} Wq(B)$$

is exact ($i: A \longrightarrow B$ is the natural inclusion). Let $s: B \longrightarrow A$ be the trace map defined by $s(1) = 0$, $s(\delta) = 1$. Then s induces a group homomorphism $s_*: Wq(B) \longrightarrow Wq(A)$ with the property $s_*(i^*(x))$

$= s_*(1)x$ for all $x \in Wq(A)$. Since

$$s_*(<1>) \cong \begin{pmatrix} 0 & 1 \\ 1 & 0 \end{pmatrix}$$

we get $s_*(i^*(x)) = 0$ for all $x \in Wq(A)$, i.e. $Im(i^*) \subseteq Ker(s_*)$. Now we want to prove that the equality actually holds. To this end let us introduce the following notation. The direct decomposition $A/r = \prod_m A/m$ determines idempotents $\bar{e}_m \in A/m$ with $\bar{e}_m \bar{e}_{m'} = 0$ for $m \neq m'$ and $1 = \sum_m \bar{e}_m$. Let $e_m \in A$ be a representative of $\bar{e}_m$ in A for every $m \in max(A)$. Then $e_m^2 \equiv e_m \pmod m$, $e_m \cdot e_{m'} \equiv 0 \pmod r$ for $m \neq m'$ and $1 \equiv \sum_m e_m \pmod r$. Now we state the main result of this section.

(5.2) Theorem. The sequence of Witt groups

$$0 \longrightarrow W(A)[1,-t] \longrightarrow Wq(A) \xrightarrow{i^*} Wq(B) \xrightarrow{s^*} Wq(A)$$

is exact.

Proof. It remains to prove $Ker(s_*) \subseteq Im(i^*)$. This fact follows immediately from the following result.

(5.3) Theorem. Let (E,q) be a quadratic space over B, such that $s_*(E)$ is hyperbolic over A. Then there exists a quadratic space (E_0,q_0) over A, such that $(E,q) \cong (E_0 \otimes B,\ q_0 \otimes B)$.

Proof. First we consider the case $\dim E = 1$. It is enough to show that for any $m \in max(A)$ there exists $x_m \in E$ with $q(x_m) \in A \setminus m$. Since, assuming this fact, we can define $x = \Sigma e_m x_m \in E$, and we get $q(x) \equiv \sum_m e_m\, q(x_m) \pmod r$, that is $q(x) \in A^*$. Now defining $E_0 = Ax$, it follows that $E \cong E_0 \otimes B$. The assumption $s_*(E) \cong [0,0]$ means that there exist $e,f \in E$, such that $E = Ae \oplus Af$, $q(e) = a$, $q(f) = b$ and $(e,f) = c + \delta$, where $a,b,c \in A$. If $a \notin m$ or $b \notin m$, we choose $x_m = e$ or $x_m = f$, respectively. Thus we may assume that $a,b \in m$. Over m there exist at most two maximal ideals M,M' of B with $M \cap M' = mB$. Let $\bar{e}$, $\bar{f}$ be the images of e, f in E/ME, respectively. Then $\bar{e} \neq 0$ or $\bar{f} \neq 0$, since $E = Ae \oplus Af$, Let us assume without restriction that $E/ME = <\bar{e}>$. Since $\bar{q}(\bar{e}) = 0$ (because $a \in m$), it follows that E/ME is singular over B/M, which is a contradiction. The same holds in case $\bar{f} \neq 0$. Thus we must have $a \notin m$ or $b \notin m$, proving our assertion in the case $\dim E = 1$.

Let us now assume that $\dim E \geq 2$. We want to show that for any $m \in \max(A)$ $x_m, y_m \in E$ exist such that

(i) $\qquad q(x_m),\ q(y_m),\ (x_m, y_m) \in A$

(ii) $\qquad 4q(x_m)\,q(y_m) - (x_m, y_m)^2 \in A^*$

Using this fact we define $x = \sum_m e_m x_m$ and $y = \sum_m e_m y_m \in E$. Then (i), (ii) imply

$$q(x),\ q(y),\ (x,y) \in A$$

$$4q(x)\,q(y) - (x,y)^2 \in A^*$$

that is $F_o = Ax \oplus Ay$ is a quadratic space over A, such that $F_o \otimes B \subset E$ is a subspace of E. Then we have $E = (F_o \otimes B) \perp (F_o \otimes B)^{\perp}$. Now we use $s_*(E) \sim 0$ and the cancellation theorem (4.3), chap. III, to deducc that $s_*(F_o \otimes B)^{\perp} \sim 0$. Hence our assertion follows by induction.

To prove (i) and (ii) we shall assume that $|A/m| \geq 7$ for all $m \in \max(A)$. Let us take $m \in \max(A)$. Since $s_*(E)$ is hyperbolic, we can find a basis $\{e_1, f_1, \ldots, e_n, f_n\}$ of E over A, such that

$$q(e_i) = a_i, \quad q(f_i) = b_i, \quad (e_i, e_j) = a_{ij}$$

$$(f_i, f_j) = b_{ij}, \quad (e_i, f_e) = c_{ie}, \quad (e_i, f_i) = d_i + \delta$$

with $a_i, b_i, d_i, a_{ij}, b_{ij}, c_{ie} \in A$ for all $i,j,e = 1,\ldots,n$ and $i \neq e$. We may assume for all $i \neq j$, $i \neq e$ that

(5.4) $\qquad 4a_i a_j - a_{ij}^2,\ 4b_i b_j - b_{ij}^2,\ 4a_i b_e - c_{ie}^2 \in m,$

since if one of the above relations is not fulfilled, say for example $4a_i a_j - a_{ij}^2 \notin m$, we choose $x_m = e_i$, $y_m = e_j$ $(i \neq j)$, proving the assertions (i), (ii). Thus we assume (5.4). Now we distinguish two cases.

1) <u>$2 \in m$</u>. From (5.4) it follows that $a_{ij}, b_{ij}, c_{ie} \in m$ for all $i,j,e \neq i$. Let $M, M' \in \max(B)$ be the maximal ideals of B over m, that is with the property $M \cap M' = mB$. Then the following three cases can

only occur:

i) For all $1 \leq i \leq n$ it holds $d_i + \delta \notin M$. Then if $\bar{d}_i, \bar{\delta}$ are the residue classes of d_i and δ in B/M, respectively, we have $\bar{d}_i + \bar{\delta} \neq 0$ in B/M for all i. In this case we have $\bar{e}_i \neq 0$ in E/ME for all i and $\{\bar{e}_1,\ldots,\bar{e}_n\}$ is a basis of E/ME over B/M, because $(\bar{e}_i,\bar{f}_j) = 0$ for all $i \neq j$, $(\bar{e}_i,\bar{f}_i) \neq 0$. But on the other hand we have $(\bar{e}_i,\bar{e}_j) = 0$ for all $1 \leq i,j \leq n$, which implies that E/ME is singular, i.e. a contradiction.

ii) It holds either $d_i + \delta \in M$ or $d_i + \delta \in M'$ for all $1 \leq i \leq n$. Then the class $\overline{d_i + \delta}$ of $d_i + \delta$ in B/mB is not a unit for all i. In particular it follows that $d_i^2 + d_i - t \in m$ for all i. Since $2 \in m$, we get $(d_i - d_j)^2 + (d_i - d_j) \in m$ for all i,j and consequently

$$d_i \equiv d_j + 1 \quad \text{or} \quad d_i \equiv d_j \pmod{m}$$

If $d_1 \equiv d_2 + 1 \pmod{m}$, we define $x_m = -e_1 + e_2$ and $y_m = f_1 + f_2 \in E$. Then $q(x_m)$, $q(y_m)$ and $(x_m,y_m) = -d_1 + d_2 + c_{21} - c_{12} \in A$. Since $c_{21}, c_{12} \in m$, we get $(x_m,y_m) \equiv 1 \pmod{m}$, that is both conditions (i) and (ii) are satisfied. Let us now assume $d_i \equiv d_j \pmod{m}$ for all i,j. We set $d = d_1$ and obtain $d \equiv d_i \pmod{m}$ for all i. Hence $(\bar{e}_i,\bar{f}_i) = \bar{d} + \bar{\delta} \in B/mB$. Taking a basis $\{\bar{x}_1,\ldots,\bar{x}_n\}$ of E/mE over B/mB (note that E is free over B), we get the following relations

$$x_e = \sum_{i=1}^{n} \bar{\alpha}_{ei}\, \bar{e}_i + \sum_{i=1}^{n} \bar{\beta}_{ei}\, \bar{f}_i$$

with $\bar{\alpha}_{ei}, \bar{\beta}_{ei} \in A/mA$, and therefore

$$(\bar{x}_e, \bar{x}_p) = \sum_{i=1}^{n} (\bar{\alpha}_{ei}\, \bar{\beta}_{pi} + \bar{\alpha}_{pi}\, \bar{\beta}_{ei})\ (\bar{d} + \bar{\delta})$$

for all $e,p = 1,\ldots,n$. Hence

$$((\bar{x}_e,\bar{x}_p)) \in (\bar{d} + \bar{\delta})\ B/mB$$

which implies that E/mE is singular over B/mB. This is a contradiction.

iii) There are i,j such that $d_i + \delta \notin M,M'$ and $d_j + \delta \in M$ (or $\in M'$). These two conditions imply $d_i^2 + d_i - t \notin m$ and $d_j^2 + d_j - t \in m$. Since $2 \in m$, we get $d_i \pm d_j \notin m$. Now we define $x_m = e_i - e_j$ and $y_m = f_i + f_j$. Then $q(x_m)$, $q(y_m)$, $(x_m,y_m) \in A$, where $(x_m,y_m) = d_i - d_j + c_{ij} - c_{ij}$. From our assumption $c_{ij} \in m$ $(i \neq j)$ we deduce that $(x_m, y_m) \in A^*$. This shows that x_m, y_m satisfy both conditions (i) and (ii). Hence the proof of our assertion (i), (ii) is complete in the case $2 \in m$. It should be noted that we never used the assumption $|A/m| \geq 7$. This condition is needed in the second case:

2) <u>$2 \notin m$</u>. We again distinguish several cases:

i) Let us assume $a_i, b_i \in m$ for all $i = 1,\dots,n$. Then (5.4) implies $a_{ij}, b_{ij}, c_{ie} \in m$ for all $i,j,e,\ i \neq e$. Let M be a maximal ideal of B over m. Then for any basis $\{\bar{x}_1,\dots,\bar{x}_n\}$ of E/ME over B/M we have

$$(\bar{x}_e, \bar{x}_p) = \sum_{i=1}^{n} (\bar{\alpha}_{ei}\ \bar{\beta}_{pi} + \bar{\alpha}_{pi}\ \bar{\beta}_{ei})\ (\bar{d}_i + \bar{\delta})$$

for some $\bar{\alpha}_{ei}, \bar{\beta}_{pi} \in A/m$. Since E/ME is non singular over B/M, it follows that $d_i + \delta \notin M$ for some i. If $d_i + \delta \notin M$ holds for all $i = 1,\dots,n$, one easily deduces that $\{\bar{e}_1,\dots,\bar{e}_n\}$ is a basis of E/ME over B/M, which is a contradiction, since $\bar{q}(\bar{e}_i) = 0$, $(\bar{e}_i, \bar{e}_j) = 0$ for all i,j. Therefore there exist $i \neq j$, such that $d_i + \delta \notin M$ but $d_j + \delta \in M$. In particular, we get $d_i \not\equiv d_j$ (mod m). Defining $x_m = e_i - e_j$, $y_m = f_i - f_j$, we obtain $q(x_m)$, $q(y_m)$, $(x_m, y_m) \in A$ and $(x_m, y_m) = d_i - d_j + c_{ij} - c_{ji} \notin m$. This proves our assertion in this case.

ii) Let us assume that either $a_i \notin m$ or $b_i \notin m$ for some $1 \leq i \leq n$. Without restriction we can suppose $a_1 \notin m$. Let $\alpha, a, b, e, d \in A$ be some elements (still undetermined) and define $\lambda = a + b\delta$, $\mu = c + d\delta$, $e = e_1 + \alpha e_2$, $f = \lambda e_1 + \lambda f_1 + f_2$. We claim that $\alpha, a, b, c \in A$ can be found, so that $q(e)$, $q(f)$, $(e,f) \in A$ and $4q(e)\,q(f) - (e,f)^2 \in A^*$ are satisfied. The requirement $q(e) \in A$ is automatically fulfilled because $\alpha \in A$. The requirements $q(f)$, $(e,f) \in A$ are equivalent with

$$a[2a_1 b + c + d(1 + d_1)] =$$
$$-[(2c + d)db_1 + b(ba_1 + dt + c_{12} + (c + d)(1 + d_1)) + db_{12}]$$

and

$$\alpha(1 + c_{21}d + a_{12b}) = -(2a_1 b + c + d(1 + d_1))$$

Let us take $b = 0$. Then we require

$$\text{(5.5)} \qquad \begin{aligned} &c + d(1 + d_1) \in A^* \\ &1 + c_{21}d \in A^* \end{aligned}$$

Under these conditions we get

$$a = - \frac{d[(2c + d)b_1 + b_{12}]}{c + d(1 + d_1)}$$

$$\alpha = - \frac{c + d(1 + d_1)}{1 + c_{21}\, d}$$

Hence for any two elements $c,d \in A$ satisfying (5.5) and with α, a as above, we have $q(e)$, $q(f)$, $(e,f) \in A$. Computing $4q(e)q(f) - (e,f)^2$ we get

$$4q(e)q(f) - (e,f) = \frac{P(c,d)}{u}$$

with $u \in A^*$ and $P(c,d) \in A[c,d]$ a quadratic polynomial in c and d over A, which is not identical zero. Now we require

(5.6) $$P(c,d) \in A^*$$

Now we use the assumption $|A/m| \geq 7$ to find $c,d \in A$ such that (5.5) and (5.6) are both satisfied. Then $x_m = e$ and $y_m = f$ have the required properties (i) and (ii).

Thus we have proved the theorem under the assumption $|A/m| \geq 7$ for all $m \in \max(A)$. We now consider the general case. To this end we use the well-known cubic Frobenius extension $C = A[X]/(X^3+6X^2-X+1) = A \oplus Ax \oplus Ax^2$ (see (2.3), chap. IV). Then $B \otimes_A C = C[Y]/(Y^2-Y+t) = C(\wp^{-1}(-t))$, and we get a commutative diagram for the natural inclusions.

$$\begin{array}{ccc} B & \xrightarrow{i'} & B \otimes_A C \\ i \uparrow & & \uparrow j' \\ A & \xrightarrow[j]{} & C \end{array}$$

Since $B \otimes_A C$ is a quadratic separable extension of C and $|C/n| \geq 7$ for all $n \in \max(C)$, we know that our assertion is true for this extension. Let $l': B \otimes_A C \longrightarrow B$ and $l: C \longrightarrow A$ be the trace maps defined by $1 \longrightarrow 1$, $x \longrightarrow 0$, $x^2 \longrightarrow 0$. We define also a trace map $s': B \otimes_A C \longrightarrow C$ by $s'(1) = 0$, $s'(\delta) = 1$. Then one easily sees that

$$l'_* \circ j'^* = i^* \circ l_* , \quad j^* \circ s_* = s'_* \circ i'^* .$$

Let now (E,q) be a quadratic space over B, such that $s_*(E)$ is hyperbolic over A. Then $j^* \circ s_*(E) = s'_* \circ i'^*(E)$ is hyperbolic over C, and since our theorem is true for $j'\colon C \longrightarrow B \otimes C$, there exists a quadratic space (F,p) over C, such that $i'^*(E) \cong j'^*(F) = F \otimes_C (B \otimes_A C)$. Applying the transfer map l'_* to this realtion, we obtain $l'_* \circ i'^*(E) \cong l'_* \circ j'^*(F)$. Now we use the Frobenius reciprocity law for (l',i') and the above relation to conclude that $E \otimes l'_*(<1>) \cong i^* \circ l_*(F)$, that is

$$l_*(F) \otimes_A B \cong E \perp r \times \mathbb{H}$$

for a suitable $r \geq 1$, since $l'_*(<1>) \cong <1> \perp \begin{pmatrix} 0 & 1 \\ 1 & \delta \end{pmatrix}$

Thus E is the anisotropic part of $l_*(F) \otimes B$, because we may assume without restriction that E is anisotropic. Now we use (4.2) (see also remark (4.13)) to show that there exists a quadratic space (E_0,q_0) over A with the property $(E,q) \cong (E_0 \otimes B,\ q_0 \otimes B)$. This concludes the proof of the theorem, and in particular of (5.2).

The next step is to extend the exact sequence in (5.2) to the right. This can be done using the following result.

(5.7) <u>Lemma</u>. In Wq(A) we have

$$\mathrm{Im}(s_*) = \mathrm{Ann}\,(<1,-1-4t>).$$

<u>Proof</u>. From $\delta^2 = \delta + t$ we get $1 + 4t = (1 + 2\delta)^2$, and therefore $<1,-1-4t> = <1,-(1 + 2\delta)^2> \cong <1,-1>$. Hence for any $[q] \in Wq(B)$ we have

$$\begin{aligned} <1,-1-4t> \otimes s_*(q) &\cong s_*(<1,-1-4t> \otimes q) \\ &\cong s_*(<1,-1> \otimes q) \\ &\sim 0 , \end{aligned}$$

that is $\mathrm{Im}(s_*) \subseteq \mathrm{Ann}(<1,-1-4t>)$. To prove the other inclusion let us first assume that $|A/m| \geq 7$ for all $m \in \max(A)$. We now use the following result, which will be proved in section 8 of this chapter (see (8.6), (8.11)):

$$\mathrm{Ann}(<1,-1-4t>) \;=\; \sum_{c} W(A)\ [1,c^2(1+4t)]$$

where c ranges over all units of A, such that $1-4c^2(1+4t) \in A^*$, and since we are assuming $|A/m| \geq 7$ for all $m \in \max(A)$, we may also require $1-2c \in A^*$ (see (8.6) in this chapter). Hence for any such c we must prove

$$[1,c^2(1+4t)] \in \mathrm{Im}(s_*).$$

This can be done as follows. For a suitable $y \in A^*$ with $1+2y \in A^*$, it follows that

$$s_*([1,y(-1+\delta)]) \sim <y> \otimes [1,-y+4ty^2]$$

We set $[1,-y+4ty^2] = <e,f>$, where $q(e) = 1$, $q(f) = -y+4ty^2$, $(e,f) = 1$. Since $<e,f> = <e, ye+f>$ and $(e, ye+f) = 1+2y$, $q(ye+f) = y^2+y-y+4ty^2 = y^2(1+4t)$, we obtain

$$[1,-y+4ty^2] \cong [1,\left(\frac{y}{1+2y}\right)^2 (1+4t)]$$

Choosing now $y = c/1-2c$, we deduce

$$s_*\left(<\frac{c}{1-2c}> \otimes [1,\left(\frac{c}{1-2c}\right)(-1+\delta)]\right) \sim [1,c^2(1+4t)]$$

which proves $[1,c^2(1+4t)] \in \mathrm{Im}(s_*)$. Now we drop the assumption $|A/m| \geq 7$ for all $m \in \max(A)$. Considering again the cubic Frobenius extension $C = A[X]/(X^3+6X^2-X+1)$ and using the same notation as during the proof of (5.3), we get the following commutative diagram

$$\begin{array}{ccc}
Wq(B \otimes_A C) & \xrightarrow{s'_*} & Wq(C) \\
j'^* \uparrow \;\; \downarrow l'_* & & j^* \uparrow \;\; \downarrow l_* \\
Wq(B) & \xrightarrow[s_*]{} & Wq(A)
\end{array}$$

Now we have proved in $Wq(C)$ that $\mathrm{Im}(s'_*) = \mathrm{Ann}(<1,-1-4t> \otimes C)$. Let us take $z \in \mathrm{Ann}(<1,-1-4t>)$ in $Wq(A)$. We have to show that $z \in \mathrm{Im}(s_*)$. Since $j^*(z) \in \mathrm{Ann}(<1,-1-4t> \otimes C) = \mathrm{Im}(s'_*)$, there is $w' \in Wq(B \otimes C)$ such that $j^*(z) = s'_*(w')$. We define $w = l'_*(w') \in Wq(B)$. Then $s_*(w) = s_* \circ l'_*(w') = l_* \circ s'_*(w') = l_* \circ j^*(z) =$

$zl_*(1) = z$, since $1_*(1) = 1$. This completes the proof of the lemma.

We summarize the results of (5.2) and (5.7) in the following

(5.8) Theorem. Let $B = A(\wp^{-1}(t))$ be a quadratic separable extension of A. Then the sequence

$$0 \longrightarrow W(A)[1,-t] \longrightarrow Wq(A) \xrightarrow{i^*} Wq(B) \xrightarrow{s_*} Wq(A) \xrightarrow{m} \langle 1,-1-4t\rangle Wq(A) \longrightarrow 0$$

is exact, where m means multiplication with $\langle 1,-1-4t\rangle \in W(A)$.

(5.9) Corollary. Let A be a semi local ring with $4 = 0$. Then for any quadratic separable extension $B = A(\wp^{-1}(t))$ of A the following sequence is exact

$$0 \longrightarrow W(A)[1,-t] \longrightarrow Wq(A) \xrightarrow{i^*} Wq(B) \xrightarrow{s_*} Wq(A) \longrightarrow 0$$

Proof. We have $\langle 1, -1-4t\rangle = \langle 1,-1\rangle$, because $4 = 0$. This implies $m = 0$. One can see directly that s_* is onto just noting that for any $a \in A$ it holds

$$s_*[-a(1+t)+a\delta,-1] \sim [1,a]$$

If $2 \in A^*$ and $B = A(\sqrt{a})$, then the sequence in (5.8) reduces to

$$0 \longrightarrow W(A)\langle\langle -a\rangle\rangle \longrightarrow W(A) \longrightarrow W(B) \longrightarrow W(A) \longrightarrow \langle\langle -a\rangle\rangle W(A) \longrightarrow 0$$

(see [L] for the field case).

(5.10) Proposition. Let A be any semi local ring and consider the quadratic separable extension $B = A(\wp^{-1}(t))$. Let q be a quadratic n-fold Pfister space over A, such that $[q] \in Im(s_*)$. Then there exists a quadratic n-fold Pfister space p over B and $d \in A^*$, such that $\langle d\rangle \otimes q \sim s_*(p)$.

Proof. For simplicity, let us assume that $|A/m| \geq 7$ for all $m \in max(A)$. Since $[q] \in Im(s_*) = Ann(\langle 1, -1-4t\rangle)$, we obtain (see (8.5), (8.6) below)

$$q \cong \varphi \otimes [1,d^2(1+4t)]$$

with a suitable (n-1)-fold bilinear Pfister space φ and $d \in A^*$, such that $1 - 2d \in A^*$. On the other hand, we have shown during the proof of (5.7) that

$$s_*(\langle c\rangle \otimes [1,c(-1+\delta)]) \sim [1,d^2(1+4t)] ,$$

where $c = d/1-2d$. Therefore

$$\begin{aligned} s_*(\varphi \otimes [1,c(-1+\delta)]) &\cong \varphi \otimes s_*([1,c(-1+\delta)]) \\ &\sim \langle c\rangle \otimes \varphi \otimes [1,d^2(1+4t)] , \end{aligned}$$

that is $s_*(p) \sim \langle c\rangle \otimes q$, where $p = \varphi \otimes [1,c(-1+\delta)]$.

§6. The torsion of Wq(A) and W(A).

As usually A denotes a semi local ring. An element $x \in Wq(A)$ or $x \in W(A)$ is called torsion element, if there exists an integer $t \geq 1$ with $tx = 0$. If (E,q) (or (M,b)) is a representative of x, we call (E,q) (resp. (M,b)) a torsion space. We shall denote the torsion subgroups of $Wq(A)$ and $W(A)$ by $Wq(A)_t$ and $W(A)_t$, respectively. An immediate consequence of (3.2), chap.IV, is the following result.

(6.1) Lemma. Let $x \in Wq(A)$ be a torsion element, which is represented by a Pfister space (E,q). Then the order of x is a power of 2.

Proof. Let $r \geq 1$ be the order of x. In particular $r \times E \sim 0$. Let t be an integer, such that $2^t \geq r$. Therefore $2^t \times E$ is isotropic, and since $2^t \times E$ is a quadratic Pfister space, we get $2^t \times E \sim 0$ (see (3.2), chap.IV). Hence $2^t x = 0$, and therefore $r \mid 2^t$. This proves the lemma.

(6.2) Lemma. Assume that $x \in Wq(A)$ is represented by a quadratic space $q = \varphi \otimes [1,b]$, where φ is a bilinear space over A. If x is a torsion element, then the order of x is a power of 2.

Proof. Without restriction we may assume that $\varphi = \langle a_1,\dots,a_n\rangle$ with $a_i \in A^*$. Now we use a construction due to Witt (see [Lo]). For any sequence of signatures $\varepsilon = (\varepsilon_1,\dots,\varepsilon_n)$, $\varepsilon_i = \pm 1$, we construct the bilinear Pfister space $\gamma(\varepsilon,\varphi) = \langle\langle \varepsilon_1 a_1,\dots,\varepsilon_n a_n\rangle\rangle$. Then $\langle a_i\rangle \otimes \gamma(\varepsilon,\varphi) \cong \langle \varepsilon_i\rangle \otimes \gamma(\varepsilon,\varphi)$ for all i, which implies that

$$\gamma(\varepsilon,\varphi) \otimes q \cong \gamma(\varepsilon,\varphi) \otimes \langle\varepsilon_1,\ldots,\varepsilon_n\rangle \otimes [1,b]$$

$$\sim \bar{\varepsilon} \times \gamma(\varepsilon,\varphi) \otimes [1,b] ,$$

where $\bar{\varepsilon} = \varepsilon_1 + \ldots + \varepsilon_n$. From this relation and the fact that q is a torsion space, we deduce that $\gamma(\varepsilon,\varphi) \otimes [1,b]$ is a torsion space, too. Thus, according to (6.1), the order of $\gamma(\varepsilon,\varphi) \otimes q$ is a power of 2. On the other hand, the following relation holds in W(A)

$$(*) \qquad 2^n = \sum_{\varepsilon} [\gamma(\varepsilon,\varphi)] ,$$

where ε runs over all sequences of signatures. This relation can easily be proved by induction. In particular we get

$$2^n \times [q] = \sum_{\varepsilon} [\gamma(\varepsilon,\varphi) \otimes q] .$$

Since the order of every element $\gamma(\varepsilon,\varphi)\otimes q$ is a power of 2, it follows from the relation above that the order of [q] is a power of 2, too.

(6.3) Theorem. The order of any torsion element in Wq(A) is a power of 2.

Proof. Let $x = [q]$ be a torsion element of Wq(A). Replacing x by 2x if necessary, we may assume that the dimension of q is even. Therefore we can write

$$q = \psi_1 \otimes [1,b_1] \perp \ldots \perp \psi_m \otimes [1,b_m] ,$$

with suitable bilinear spaces ψ_i. Now we proceed by induction on m. The case $m = 1$ was treated in lemma (6.2). Let us assume that our assertion is true for all semi local rings and all torsion spaces, which admit a representation as above of lenght < m. Over the extension $B = A(\wp^{-1}(-b_m))$ we get

$$q \otimes B \sim \psi_1 \otimes [1,b_1] \perp \ldots \perp \psi_{m-1} \otimes [1,b_{m-1}] .$$

By induction, it follows that $2^r \times (q \otimes B) \sim 0$ over B for a suitable $r \geq 0$. Using (4.11) we deduce that $2^r \times q \sim \varphi \otimes [1,b_m]$, where φ is a bilinear space over A. In particular, it follows that $\varphi \otimes [1,b_m]$ is a torsion space, whose order is a power of 2 in virtue of (6.2). This proves that the order of [q] is a power of 2, proving the theorem.

(6.4) Corollary. Let q and φ be a quadratic and bilinear space over A, respectively. Assume that $\varphi \otimes q \sim 0$. If q is not hyperbolic, then $\dim \varphi$ is even.

Proof. (Pfister). We may assume that $\varphi = \langle a_1, \dots, a_n \rangle$, $a_i \in A^*$. Assuming that our assertion is false, we can choose n odd and minimal, such that $\varphi \otimes q \sim 0$ for a quadratic space $q \not\sim 0$. Of course we can take $a_1 = 1$. Then for all $2 \leq i \leq n$ it holds that $\langle 1, -a_i \rangle \otimes \varphi \otimes q \sim 0$ and therefore

$$\langle a_2, \dots, a_{i-1}, a_{i+1}, \dots, a_n \rangle \otimes \langle 1, -a_i \rangle \otimes q \sim 0.$$

Since n is minimal, we get $\langle 1, -a_i \rangle \otimes q \sim 0$ for all $2 \leq i \leq n$, that is $q \cong \langle a_i \rangle \otimes q$ (use (4.3), chap. III). Therefore $n \times q \cong \varphi \otimes q \sim 0$, which is a contradiction. This proves the corollary.

(6.5) Remark. 1. For a generalization of the above results we refer the reader to (7.11) in the next section.

2. Theorems (6.3) and (6.4) are generalizations of well-known results of Pfister in the field case (see $[Pf]_2$, $[Pf]_3$). They can also be proved for the Witt ring W(A) of bilinear spaces over A (see $[K]_3$ and $[K\text{-}R\text{-}W]_2$ for a different treatment using commutative algebra). We shall now deduce from (6.3) the corresponding result for W(A). To this end, we need some general remarks. Let C be a commutative ring (with 1), such that Wq(C) has only 2-torsion. Then we claim that W(C) has only 2-torsion, too. To prove this, we consider the homomorphism $\beta: Wq(C) \longrightarrow W(C)$, given by $\beta([q]) = [b_q]$ (see § 4, chap. I). Let $[b] \in W(C)$ be a torsion element. We remind that the quadratic space E_8 (see (4.13), chap. I or [Se], s. 89, or [M-H], § 6) is non singular over $\mathbb{Z}$, and hence also over C. One easily sees that $\beta([E_8]) = 8$ in $W(C)$, and therefore $\beta([b \otimes E_8]) = 8 \times [b]$. Since $[b \otimes E_8]$ is a torsion element in Wq(C), we can find by assumption an integer $r \geq 0$, such that $2^r \times [b \otimes E_8] = 0$. This implies $2^{r+3} \times [b] = 0$ in $W(C)$, proving the claim. Applying the remark above, we get from (6.3), (6.4).

(6.6) Theorem. Let A be a semi local ring. Then W(A) has only 2-torsion. More precisely, in W(A) a divisor of zero must have even dimension.

Let us consider again the commutative ring C. Then we have proved one part of the following

(6.7) Proposition. The assertions

(i) $Wq(C)$ has only 2-torsion

(ii) $W(C)$ has only 2-torsion

are equivalent.

Proof. We have proved (i) $\Longrightarrow$ (ii). Let us now assume (ii). If $[q] \in Wq(C)$ is a torsion element, then $\beta([q]) = [b_q] \in W(C)$ is torsion element, too, and therefore $\beta(2^r \times [q]) = 2^r \times [b_q] = 0$ for some $r \geq 0$. Our assertion follows now from the following

(6.8) Lemma. $8 \times \mathrm{Ker}(\beta) = 0$.

Proof. From $b_{E_8} \sim 8 \times \langle 1 \rangle$ we conclude that $b_q \otimes E_8 \cong b_{E_8} \otimes q \sim 8 \times q$ for any quadratic space q over C. If $[q] \in \mathrm{Ker}(\beta)$, then it follows that $8 \times [q] = [b_q \otimes E_8] = \beta([q])[E_8] = 0$, which proves the lemma.

As a consequence of (6.4) and (6.6) we obtain

(6.9) Theorem. Let $i: A \longrightarrow B$ be a Frobenius extension of odd degree. Then the induced homomorphisms $i^*: Wq(A) \longrightarrow Wq(B)$ and $i^*: W(A) \longrightarrow W(B)$ are both one to one.

Proof. Let $s: B \longrightarrow A$ be a trace map. Consider $y \in Wq(A)$ or $y \in W(A)$. Then we know that $s_*(i^*(y)) = s_*(1)y$. If $y \in \mathrm{Ker}(i^*)$, it follows that $s_*(1)y = 0$, and since $s_*(1)$ is represented by a space of odd dimension, we must have $y = 0$, proving the assertion.

§ 7. The local global principle of Pfister.

Let k be a field of characteristic $\neq 2$. Using the orderings of k Pfister described in his well-known paper $[Pf]_3$ (Satz 22) the torsion part $W(k)_t$ of $W(k)$ in the following manner: for each ordering α of k let k_α be the real closure of k relative to α. Then the inclusions $k \longrightarrow k_\alpha$ define a homomorphism $i^*: W(k) \longrightarrow \prod_\alpha W(k_\alpha)$ and $W(k)_t = \mathrm{Ker}(i^*)$. If k is non real, then of course the above result means the fact $W(k) = W(k)_t$. Let us assume that k is formally real. Then for any real closure k_α of k we have $W(k_\alpha) \cong \mathbb{Z}$ by Silvester's law of inertia, and the induced homomorphism $W(k) \longrightarrow W(k_\alpha)$ is simply a ring homomorphism $W(k) \longrightarrow \mathbb{Z}$. Conversely, it can be shown that every ring homomorphism $W(k) \longrightarrow \mathbb{Z}$ corresponds to an ordering α of the field k and it can be recovered by $W(k) \longrightarrow W(k_\alpha) \cong \mathbb{Z}$. Thus there exists a one to one correspondence between orderings of the field k and ring homomorphisms $W(k) \longrightarrow \mathbb{Z}$ (see for example [L] or [Lo]). Generalizing this ideas for commutative rings, Knebusch introduced the concept of signature of a commutative ring A simply as a ring homomorphism $W(A) \longrightarrow \mathbb{Z}$, though this concept do not has apparently an interpretation as an ordering of the ring A (see $[K]_5$, $[K]_6$ for details). But if A is a semi local ring, it can be shown that the signatures of A (in the above sense) correspond to the group homomorphisms $\sigma: A^* \longrightarrow \{\pm 1\}$ with the properties: i) $\sigma(-1) = -1$, ii) $\sigma(a) = 1$ for every $a \in A^*$ of the form $a = c_1a_1 + \ldots + c_ra_r$, where $\sigma(a_i) = 1$ and c_i is a sum of squares in A, $1 \leq i \leq r$. This correspondence can be seen as follows. Let $\sigma: W(A) \longrightarrow \mathbb{Z}$ be a signature of A. Then we define a group homomorphism $\bar{\sigma}: A^* \longrightarrow \{\pm 1\}$ by $\bar{\sigma}(a) = \sigma(<a>)$ for all $a \in A^*$. We briefly show that $\bar{\sigma}$ has the above properties (i) and (ii). Since $<1> \perp <-1> \sim 0$ over A, we get $\bar{\sigma}(-1) = \sigma(<-1>) = \sigma(-<1>) = -1$. To show (ii) we set N for the maximal number of squares appearing in the c_i's. Then a is represented by the bilinear space $N \times <a_1> \perp \ldots \perp N \times <a_r> = b$. Hence $b \cong <a> \perp b_0$ and therefore $<1> \perp b \cong <a> \perp (<1> \perp b_0)$. Now, $<1> \perp b_0$ is a proper bilinear space, so that it admits an orthogonal basis. Thus $<1> \perp b \cong <a> \perp <b_1,\ldots,b_s>$. Comparing the signatures on both sides, we get $\sigma(<a>) + \sum_{i=1}^{s} \sigma(<b_i>) = 1 + Nr$, and since $s = Nr$, we must have $\bar{\sigma}(a) = \sigma(<a>) = 1$. This proves (ii). Conversely, any homomorphism $\bar{\sigma} : A^* \longrightarrow \{\pm 1\}$ with the properties (i) and (ii) induces a ring homomorphism $\sigma: W(A) \longrightarrow \mathbb{Z}$ by $\sigma([<a_1,\ldots,a_n>]) = \bar{\sigma}(a_1) + \ldots + \bar{\sigma}(a_n)$. Using this notion of signature the local global principle of Pfister has been extended to semi local rings in $[K\text{-}R\text{-}W]_{1,2}$ in the following way: if A is semi local, then $W(A)_t = \operatorname{Ker} [W(A) \longrightarrow \prod_\sigma \mathbb{Z}]$, where σ runs over all signatures of A (see (7.16) for the proof). In this section we shall prove this theorem and the analogue for Wq(A). We now give formally the important.

(7.1) Definition. Let A be a semi local ring. A signature of A is a ring homomorphism $\sigma : W(A) \longrightarrow \mathbb{Z}$. We denote the set of signatures of A by Sig(A). If Sig(A) $\neq \emptyset$, then A is called formally real (or simply: real), otherwise A is called non real (or imaginary).

In (7.8) below we shall characterize the non real semi local rings.

Using the homomorphism β: $Wq(A) \longrightarrow W(A)$ we can associate to every signature σ: $W(A) \longrightarrow \mathbb{Z}$ (if it exists) a non zero ring homomorphism $\bar{\sigma} = \sigma \circ \beta : Wq(A) \longrightarrow \mathbb{Z}$ (compare (7.6) below). Since for any $x \in W(A)$, $y \in Wq(A)$ it holds that $\bar{\sigma}(xy) = \sigma(x)\bar{\sigma}(y)$, we see that $Ker(\bar{\sigma})$ is a $W(A)$-submodule of $Wq(A)$. Conversely, we want to show that every non zero homomorphism (of rings) $\bar{\sigma}$: $Wq(A) \longrightarrow \mathbb{Z}$, whose kernel is a $W(A)$-submodule of $Wq(A)$, has the form $\bar{\sigma} = \sigma \circ \beta$ for some $\sigma \in Sig(A)$. We also show that the correspondence $\sigma \longleftrightarrow \bar{\sigma}$ is one to one. Suppose that $\bar{\sigma}$: $Wq(A) \longrightarrow \mathbb{Z}$ is a ring homomorphism as above. Then $\bar{\sigma}[Wq(A)] = \mathbb{Z}\, n$ for some integer $n \geq 1$ (since $\bar{\sigma}$ is assumed to be non zero). Take $[q_0] \in Wq(A)$ with $\bar{\sigma}([q_0]) = n$. Then for any $[b] \in W(A)$ we define

$$\text{(7.2)} \qquad \sigma([b]) = \frac{\bar{\sigma}([b \otimes q_0])}{n} \in \mathbb{Z}$$

We assert that σ: $W(A) \longrightarrow \mathbb{Z}$ is a well-defined ring homomorphism, such that $\bar{\sigma} = \sigma \circ \beta$. This follows in several steps.

(7.3) For any $[q] \in Wq(A)$ with $\bar{\sigma}([q]) \neq 0$ and every $[b] \in W(A)$ we have $\bar{\sigma}([b \otimes q])/\bar{\sigma}([q]) = \bar{\sigma}([b \otimes q_0])/\bar{\sigma}([q_0])$. This follows immediately from the relation $(b \otimes q) \circ q_0 \cong q \circ (b \otimes q_0)$, because $\bar{\sigma}$ is a ring homomorphism and hence $\bar{\sigma}([b \otimes q])\ \bar{\sigma}([q]) \cong \bar{\sigma}([q])\ \bar{\sigma}([b \otimes q_0])$ Thus, the definition (7.2) does not depend on the choise of q_0.

(7.4) $\sigma : W(A) \longrightarrow \mathbb{Z}$ is a ring homomorphism. Take $[b_1]$, $[b_2] \in W(A)$. First suppose that $\sigma([b_1])\sigma([b_2]) \neq 0$. In particular $\bar{\sigma}([b_2 \otimes q_0]) \neq 0$ and hence (see (7.3))

$$\sigma([b_1][b_2]) = \sigma([b_1 \otimes b_2]) = \frac{\bar{\sigma}([b_1 \otimes b_2 \otimes q_0])}{\bar{\sigma}([q_0])}$$

$$= \frac{\bar{\sigma}([b_1 \otimes b_2 \otimes q_0])}{\bar{\sigma}([b_2 \otimes q_0])} \cdot \frac{\bar{\sigma}([b_2 \otimes q_0])}{\bar{\sigma}([q_0])}$$

$$= \sigma([b_1])\ \sigma([b_2]).$$

If, for example, $\sigma([b_2]) = 0$, i.e. $\bar{\sigma}([b_2 \otimes q_0]) = 0$, then $[b_2 \otimes q_0] \in Ker(\bar{\sigma})$ and hence $[b_1][b_2 \otimes q_0] = [b_1 \otimes b_2 \otimes q_0] \in Ker(\bar{\sigma})$, since $Ker(\bar{\sigma})$ is a $W(A)$-submodule. Therefore $\sigma([b_1][b_2]) = 0 = \sigma([b_1])\sigma([b_2])$. This proves the claim.

(7.5) It holds $\bar{\sigma} = \sigma \circ \beta$. Because for every $[q] \in Wq(A)$

$$\sigma \circ \beta([q]) = \sigma([b_q]) = \frac{\bar{\sigma}([b_q \otimes q_o])}{\bar{\sigma}([q_o])} = \frac{\bar{\sigma}([q \circ q_o])}{\bar{\sigma}([q_o])} = \bar{\sigma}([q])$$

(7.6) Let $\sigma: W(A) \longrightarrow \mathbb{Z}$ be a signature of A. Then $\sigma \circ \beta \neq 0$. If $\sigma \circ \beta = 0$, then for every $[b] \in W(A)$ we have $\sigma \circ \beta([b \otimes E_8]) = \sigma(8[b]) = 8\sigma([b]) = 0$, that is $\sigma([b]) = 0$, which is a contradiction.

Let $\overline{Sig}(A)$ be the set of non zero ring homomorphisms $\sigma : Wq(A) \longrightarrow \mathbb{Z}$, whose kernel is a W(A)-submodule of Wq(A) (if they exist). Then (7.2) defines a map $\overline{Sig}(A) \longrightarrow Sig(A)$, which is obviously the inverse of the map $Sig(A) \longrightarrow \overline{Sig}(A)$, $\sigma \longrightarrow \sigma \circ \beta$. Thus we have a one to one correspondence between signatures of A and the elements of $\overline{Sig}(A)$. Accordingly, we call the elements of $\overline{Sig}(A)$ signatures, too.

(7.7) Lemma. The following assertions are equivalent.

i) $W(A)_t = \bigcap_{\sigma \in Sig(A)} Ker(\sigma)$

ii) $Wq(A)_t = \bigcap_{\bar{\sigma} \in \overline{Sig}(A)} Ker(\bar{\sigma})$

Proof. (i) $\Longrightarrow$ (ii). Clearly $Wq(A)_t \subseteq \cap Ker(\bar{\sigma})$. Conversely, for any $[q] \in \cap Ker(\bar{\sigma})$ we have $[b_q] \in \cap Ker(\sigma) = W(A)_t$, and therefore $[q] \in Wq(A)_t$ (see (6.8)).

(ii) $\Longrightarrow$ (i). Take $[b] \in \cap Ker(\sigma)$. Hence $[b \otimes E_8] \in \cap Ker(\bar{\sigma})$, since $\bar{\sigma}([b \otimes E_8]) = \sigma([b])\, \bar{\sigma}([E_8]) = 0$. Therefore $[b \otimes E_8] \in Wq(A)_t$ and consequently $\beta([b \otimes E_8]) = 8[b] \in W(A)_t$, that is $[b] \in W(A)_t$.

Now we prove the assertion (i) in (7.7) following $[K\text{-}R\text{-}W]_{1,2}$. First, let us consider the special case $Sig(A) = \emptyset$ (i.e. $\overline{Sig}(A) = \emptyset$).

(7.8) Theorem. For any semi local ring A the following assertion are equivalent.

i) $Sig(A) = \emptyset$

ii) -1 is a sum of squares in A.

Proof. (ii) $\Longrightarrow$ (i). Let us assume $-1 = a_1^2 + \ldots + a_s^2$, $a_i \in A$, and take $2^t \geq s + 1$ for some integer $t \geq 1$. Hence the bilinear Pfister space $2^t \times \langle 1 \rangle$ is isotropic over A, and consequently (see

(3.5), chap. IV) $2^{t+1} \times \langle 1\rangle \sim 0$. If $\mathrm{Sig}(A) \neq \emptyset$, then

$$\sigma([2^{t+1} \times \langle 1\rangle]) = 2^{t+1}\,\sigma(1) = 0$$

i.e. $\sigma(1) = 0$ for $\sigma \in \mathrm{Sig}(A)$, which is absurd. Thus $\mathrm{Sig}(A) = \emptyset$.

(i) $\Longrightarrow$ (ii). Let us assume that $\mathrm{Sig}(A) = \emptyset$. Consider the ring homomorphism $\mathbb{Z} \longrightarrow W(A)$ given by $n \longrightarrow n\times[\langle 1\rangle]$. Then this homomorphism is not injective. Because on the contrary, there would exists a prime ideal $p \subset W(A)$, such that $p \cap \mathbb{Z} = \{0\}$, since $\mathbb{Z} \longrightarrow W(A)$ is an entire extension (see (7.12) below) and we can apply the "going up". Therefore we would have an inclusion $\mathbb{Z} \longrightarrow W(A)/p$. But on the other hand, $W(A)/p$ is generated over $\mathbb{Z}$ by the classes $\langle \bar{a}\rangle$, $a \in A^*$, with $\langle \bar{a}\rangle^2 = 1$. Since $W(A)/p$ has no zero-divisors, we have $\langle \bar{a}\rangle = \pm 1$, that is $W(A)/p = \mathbb{Z}$. Hence we get a signature $\sigma\colon W(A) \longrightarrow W(A)/p = \mathbb{Z}$, which contradics our assumption (i). This proves that $\mathbb{Z} \longrightarrow W(A)$ can not be injective. Therefore there is an integer $n \geq 1$ with $n\times\langle 1\rangle \sim 0$. Now our assertion follows from the following.

(7.9) Lemma. If there exists $n \geq 1$, such that $n\times\langle 1\rangle \sim 0$ over A, then -1 is a sum of squares in A.

Proof. Let us choose an integer $h \geq 1$, such that $1 - 4h \in A^*$. Consider the quadratic space $[1,h]$ over A. Choosing $2^t \geq n$, we see that $n\times[1,h]$ is a subspace of $2^t\times[1,h]$. Now $n\times\langle 1\rangle \sim 0$ implies $n\times[1,h] \sim 0$ and hence $n\times[1,h] \cong n\times\mathbb{H}$, thus $2^t\times[1,h]$ is isotropic. Therefore $2^t\times[1,h] \sim 0$ (see (3.2), chap. IV). Using again the cancellation theorem (see (4.3), chap. III) we get

$$2^{t-1} \times [1,h] = -2^{t-1} \times [1,h]$$

and hence $-1 \in \underline{D}(2^{t-1} \times [1,h])$. Hence there exist $a_i, b_i \in A$, such that $-1 = \sum_i (a_i^2 + a_ib_i + hb_i^2)$. Multiplying this equation by 2, we get

$$-1 = 1 + \sum_i 2(a_i^2 + a_ib_i + hb_i^2)$$

$$-1 = 1 + \sum_i [(a_i + b_i)^2 + (2h - 1)b_i^2],$$

that is -1 is a sum of squares in A. The proof of (7.9), and hence of (7.8), is complete.

Proof of the assertion (i) in (7.7). First we assume that $\mathrm{Sig}(A) = \emptyset$.

Then we have shown (see proof of (7.8)) that <1> is a torsion space, that is $W(A) = W(A)_t$. Thus the assertion (i) in (7.7) is clear. From now on we shall assume that $Sig(A) \neq \emptyset$. Under this assumption we can prove the following facts.

(7.10) The map $\mathbb{Z} \longrightarrow W(A)$ given by $n \longrightarrow n \times [<1>]$ is one to one. This follows immediately from (7.8) and (7.9), since $Sig(A) \neq \emptyset$.

(7.11) $W(A)_t = Nil(W(A))$.
This follows from (8.9) (see next section), because our assumption $Sig(A) \neq \emptyset$ implies $W(A)_t \subseteq I$ (see (6.6) and (7.10)), and therefore $Nil(W(A)) = I \cap W(A)_t = W(A)_t$.

(7.12) The extension $\mathbb{Z} \longrightarrow W(A)$ is integral. In particular the "going-up" and "lying-over" theorems are valid (see [Z-S]). This follows from the fact that $W(A)$ is generated over $\mathbb{Z}$ by the elements $[<a>]$ with the relations $[<a>]^2 = 1$ for all $a \in A^*$.

(7.13) A prime ideal $\underline{p} \subset W(A)$ is minimal if and only if $\underline{p} \cap \mathbb{Z} = \{0\}$.

<u>Proof</u>. Let $\underline{p} \subset W(A)$ be a minimal prime ideal. If $\underline{p} \cap \mathbb{Z} = \mathbb{Z}p$ for some prime number p, then $\underline{p}$ must be maximal, because of the "lying-over" theorem. Therefore every element of the unique prime ideal $\underline{p}W(A)_{\underline{p}}$ of $W(A)_{\underline{p}}$ is nilpotent. Since $p \in \underline{p}$, we get $p^n\, 1 = 0$ in $W(A)_{\underline{p}}$ for some $n \geq 1$. Therefore there exists $z \in W(A) \setminus \underline{p}$, such that $p^n z = 0$. In particular $z \in W(A)_t = Nil(W(A)) \subset \underline{p}$, which is a contradiction. Thus, we have proved $\underline{p} \cap \mathbb{Z} = \{0\}$ for every minimal prime ideal $\underline{p} \subset W(A)$. Conversely, let $\underline{p} \subset W(A)$ be a prime ideal, such that $\underline{p} \cap \mathbb{Z} = \{0\}$. Let p be any prime number. Applying the "lying-over" to $\underline{p} \cap \mathbb{Z} = \{0\} \subset p\,\mathbb{Z}$, we get a prime ideal $m \subset W(A)$ with $\underline{p} \subset m$ and $m \cap \mathbb{Z} = p\,\mathbb{Z}$. In particular $\underline{p} \neq m$. Since $\dim W(A) = 1$ (see (7.12)), we see that $\underline{p}$ is minimal and m is maximal. We also conclude that $m = \underline{p} + pW(A)$ and m is the unique maximal ideal in $W(A)$ containing $\underline{p}$, such that $m \cap \mathbb{Z} = p\,\mathbb{Z}$.

In particular we have shown

(7.14) Let $\underline{p} \subset W(A)$ be a minimal prime ideal. Then the set $\{\underline{p} + pW(A) \mid p$ prime number$\}$ is the totality of maximal ideals of $W(A)$ containing $\underline{p}$.

(7.15) The minimal prime ideals of W(A) are exactly the kernels of all signatures of A.

Proof. If $\sigma \in \mathrm{Sig}(A)$, then clearly $\mathrm{Ker}(\sigma)$ is a minimal prime ideal of W(A), since $W(A)/\mathrm{Ker}(\sigma) \cong \mathbb{Z}$. Now suppose that $p \subset W(A)$ is any minimal prime ideal. Then $p \cap \mathbb{Z} = \{0\}$ (see (7.13)), and hence $\mathbb{Z} \longrightarrow W(A)/p$ is one to one. Now $W(A)/p$ is an integral domain generated over $\mathbb{Z}$ by elements $\overline{<a>}$ with $\overline{<a>}^2 = 1$, $a \in A^*$. Hence $\overline{<a>} = \pm 1$ and $W(A)/p = \mathbb{Z}$. Thus, the canonical homomorphism σ : $W(A) \longrightarrow W(A)/p = \mathbb{Z}$ is a signature and $\mathrm{Ker}(\sigma) = p$. This proves (7.15).

The assertion (i) in (7.7) follows now at once. Namely $W(A)_t = \mathrm{Nil}(W(A)) = \cap\, p$, where p runs over all minimal prime ideals of W(A), hence according to (7.15) $W(A)_t = \cap\, \mathrm{Ker}(\sigma)$, $\sigma \in \mathrm{Sig}(A)$. Thus we have proved

(7.16) Theorem. For any semi local ring A it holds

$$W(A)_t = \bigcap_{\sigma \in \mathrm{Sig}(A)} \mathrm{Ker}(\sigma)$$

$$Wq(A)_t = \bigcap_{\overline{\sigma} \in \overline{\mathrm{Sig}}(A)} \mathrm{Ker}(\overline{\sigma})$$

(7.17) Remark. A close look at the above proof of (7.16) shows that we never used that A is semi local, but only the fact the bilinear spaces over A have a diagonal form, i.e. W(A) is generated by the classes [<a>], $a \in A^*$. Thus if Wd(A) denotes the subring of W(A) generated by the classes [<a>], $a \in A^*$, where A is any commutative ring (with 1), then (7.16) could be true for Wd(A). In this case a signature of A is a ring homomorphism σ: $Wd(A) \longrightarrow \mathbb{Z}$ (compare (8.9) and (8.11)).

To give a full description of the set of prime ideals of W(A) we must still consider the case $\mathrm{Sig}(A) = \emptyset$. In this case the situation is very simple, because $W(A) = W(A)_t$, so that $I = \mathrm{Nil}(W(A))$ (see (8.9)). This implies that I is the unique prime ideal of W(A). Now we collect all these results on prime ideals of W(A) in the following

(7.18) Theorem. Let A be a semi local ring.

(i) If $\mathrm{Sig}(A) = \emptyset$, then I is the unique prime ideal of W(A).

(ii) If $\mathrm{Sig}(A) \neq \emptyset$, then the prime ideals of $W(A)$ are either minimal or maximal. The minimal prime ideals are exactly the kernels of the signatures of A. For any minimal prime ideal $\underline{p}$ of $W(A)$ the set $\{\underline{p} + pW(A) \mid p$ prime number$\}$ is the totality of maximal ideals containing $\underline{p}$. I is the unique maximal ideal containing 2.

For further reading we refer to $[K]_{5,6,7}$.
Let us now make some applications of these results. The following proposition generalizes a well-known result of Pfister (see $[Pf]_3$).

(7.19) Proposition. Let $a \in A^*$ be a unit with $\sigma(a) = 1$ for all $\sigma \in \mathrm{Sig}(A)$. Then a has the form

$$a = \sum_{i=1}^{r} (a_i^{\,2} + a_ib_i + hb_i^{\,2})$$

$(a_i, b_i \in A)$, where h is a positive integer with $1-4h \in A^*$. In particular, if $2 \in A^*$, then a is a sum of squares.

Proof. Let us choose a positive integer h with $1-4h \in A^*$ and define the quadratic space $[1,h]$. Since $\sigma(<1,-a>) = 0$ for all $\sigma \in \mathrm{Sig}(A)$, we deduce from (7.16) that $r \times <1,-a> \sim 0$ for some $r \geq 1$. Therefore $r \times <1,-a> \otimes [1,h] \sim 0$, so that $r \times [1,h] \cong r \times <a> \otimes [1,h]$ by the cancelation theorem (see (4.3), chap. III). In particular, $r \times [1,h]$ represents a, proving the first assertion. If $2 \in A^*$, then we have $a^2 + ab + hb^2 = (a + \frac{1}{2}b)^2 + (4h - 1)(\frac{1}{2}b)^2$, proving the second assertion. It should be noted that according to (5.2), chap. III, we can write, in case $|A/m| \geq 3$ for all $m \in \max(A)$,
$a = \Sigma\, c_i^{\,2}(1 + d_i + hd_i^{\,2})$ with c_i, $1 + d_i + hd_i^{\,2} \in A^*$.

(7.20) Remark. Let $h \in \mathbb{N}$ be an integer with $1-4h \in A^*$. Then for every $a \in \underline{D}([1,h])^*$ and any $\sigma \in \mathrm{Sig}(A)$ (if $\neq \emptyset$) it holds that $\sigma(<a>) = 1$. To see this, let us write $a = b^2 + bc + hc^2$ for some $b,c \in A$. Hence $2 \times <1,-a>$ represents $-2a = -b^2 - (b+c)^2 - (2h-1)c^2$, so that $2^t \times <1,-a>$ is isotropic for some $t \geq 1$. Then (3.5), chap. IV, implies that $2^{t+1} \times <1,-a> \sim 0$. Hence for every $\sigma \in \mathrm{Sig}(A)$ we have $\sigma(<1,-a>) = 0$, i.e. $\sigma(<a>) = 1$. A more general result is the following: if $a_1,\ldots,a_r \in A^*$ are units, such that $\sigma(<a_i>) = 1$ for some $\sigma \in \mathrm{Sig}(A)$, $1 \leq i \leq r$, then $\sigma(<a>) = 1$ for any $a \in \underline{D}(<a_1,\ldots,a_r> \otimes [1,h])^*$. Applying this fact to $a_1 = \ldots = a_r = 1$, we see that the converse of (7.19) holds, too.

As a second application of (7.16) we shall shortly study the properties SAP and WAP which where first introduced in $[K-R-W]_1$. Let us consider a real semi local ring A. For any $a \in A^*$ we define $W(a) = \{\sigma \in Sig(A) \mid \sigma(<a>) = -1\}$. Then it is easy to see that the sets $\{W(a) \mid a \in A^*\}$ form a subbasis for a topology on Sig(A). With this topology Sig(A) is compact, totally disconnected and Hausdorff. Now we define the following properties for this topology:

SAP (strong approximation property): every open and closed subset of Sig(A) has the form W(a) for some $a \in A^*$.

WAP (weak approximation property): the set $\{W(a) \mid a \in A^*\}$ is a basis for the topology of Sig(A).

Over a real field these two properties are equivalent (see $[E-L]_5$). Rosenberg and Ware have recently proved the equivalence of SAP and WAP for semi local rings where 2 is a unit using elementary methods (see [R-W]). Our next aim is to show the equivalence of SAP and WAP for any semi local ring A using similar methods. We first remark that WAP and SAP are a consequence of the following property of A.

(*) for any set of units $a_1,\dots,a_m \in A^*$ there exists $a \in A^*$ such that $W(a_1) \cap \dots \cap W(a_n) = W(a)$.

Of course WAP follows immediately from (*). Let us now show (*) $\Longrightarrow$ SAP. Let Y be a closed and open subset of Sig(A). Since $Sig(A) \setminus Y$ is closed, it is compact and hence $Sig(A) \setminus Y = W(-a_1) \cup \dots \cup W(-a_n)$ for some $a_i \in A^*$ (here we use WAP). Taking complements we obtain $Y = \bigcap_i [Sig(A) \setminus W(-a_i)] = \bigcap_i W(a_i) = W(a)$ for a suitable $a \in A^*$. This proves SAP.

(7.21) Theorem. For any semi local ring SAP, WAP and (*) are equivalent.

Proof. Obviously SAP implies WAP. Let us now assume WAP. We only need to show that WAP implies (*). Take $a_1,\dots,a_n \in A^*$. We choose $h \in \mathbb{N}$ with $1-4h, 1-2h \in A^*$. Since the sets $\{W(a) \mid a \in A^*\}$ are a basis and $W(a_1) \cap \dots \cap W(a_n)$ is compact, we can find $b_1,\dots, b_m \in A^*$ such that

$$W(a_1) \cap \dots \cap W(a_n) = W(b_1) \cup \dots \cup W(b_m).$$

Repeating some intersections or unions in the above relation if necessary we see that we can assume $n = m$, i.e. $W(a_1) \cap \ldots \cap W(a_n) = W(b_1) \cup \ldots \cup W(b_n)$. Now we define

$$E = \langle\langle -a_1,\ldots,-a_n,h]] \quad \text{and} \quad F = \langle\langle b_1,\ldots,b_n,h]]$$

Since for any $\sigma \in \bigcap_{i=1}^{n} W(a_i)$ we have $\sigma(\langle a_i\rangle) = -1$, i.e. $\sigma(\langle -a_i\rangle) = 1$ for $1 \leq i \leq n$ and $\sigma(\langle b_j\rangle) = -1$ for some j, it follows that

$$\bar{\sigma}([E]) = \sigma(\langle\langle -a_1,\ldots,-a_n\rangle\rangle)\ \bar{\sigma}([1,h]) = 2^{n+1}$$

and $\bar{\sigma}([F]) = 0$.

Similarly, for any $\tau \in \mathrm{Sig}(A) \setminus \bigcap_{i=1}^{n} W(a_i)$ we get

$$\bar{\tau}([E]) = 0 \quad \text{and} \quad \bar{\tau}([F]) = 2^{n+1}$$

The used fact $\bar{\rho}([1,h]) = 2$ for all $\bar{\rho} \in \overline{\mathrm{Sig}}(A)$ follows from the isomorphism

$$\begin{pmatrix} 2 & 1 \\ 1 & 2h \end{pmatrix} \perp \langle -1\rangle \cong \langle 1,\ 2h-1,\ (2h-1)(1-4h)\rangle$$

Hence for all $\bar{\rho} \in \overline{\mathrm{Sig}}(A)$ we conclude that

$$\bar{\rho}([E \perp F]) = 2^{n+1}$$

$$\bar{\rho}(2^n \times [1,h]) = 2^{n+1}$$

Now we deduce from (7.16) that $[E \perp F] - [2^n \times [1,h]] \in Wq(A)_t$, that is $2^r \times (E \perp F) \sim 2^{r+n} \times [1,h]$ for some $r \geq 0$ (see (6.3)). Using the cancellation theorem (4.3), chap. III, we get

$$2^r \times E \perp 2^r \times F \cong 2^{r+n} \times [1,h] \perp 2^{r+n} \times \mathbb{H}$$

and therefore $2^r \times E \perp 2^r \times F$ is isotropic. Now (5.2) (iii), chap. III, implies that there exists $a \in A^*$ with $-a \in \underline{D}(2^r \times E)^*$ and $a \in \underline{D}(2^r \times F)^*$. Therefore $\sigma(\langle -a\rangle) = 1$ for all $\sigma \in W(a_1) \cap \ldots \cap W(a_n)$ (see (7.20)), that is $\sigma(\langle a\rangle) = -1$. Similarly, we get $\tau(\langle a\rangle) = 1$ for all $\tau \in \mathrm{Sig}(A) \setminus \bigcap_{i=1}^{n} W(a_i)$ (use again (7.20)), because $a \in \underline{D}(2^r \times F)$ and $\tau \in \bigcap_{i=1}^{n} W(-b_i)$. This facts imply obviously $W(a) = W(a_1) \cap \ldots \cap W(a_n)$, proving the theorem.

§ 8 Nilpotent elements in Wq(A) and W(A)

The purpose of this section is to describe the nilpotent elements in Wq(A), where A is any semi local ring. We remember that the product of $[q_1], [q_2] \in Wq(A)$ is defined by $[q_1] \circ [q_2] = [b_{q_1} \otimes q_2] = [b_{q_2} \otimes q_1]$. For any $[q] \in Wq(A)$ and $n \geq 1$ we write $[q]^n$ instead of $[q] \circ \ldots \circ [q]$ (n-times), and $[q]$ is nilpotent if and only if $[q]^n = 0$ for some $n \geq 1$.

(8.1) Lemma. For $q_1 = [1,a]$ and $q_2 = [1,b]$ $(1-4a,\ 1-4b \in A^*)$ and $c \in \underline{D}(q_1 \circ q_2)^*$ it holds that

$$[1,a] \circ [1,b] \cong \langle c\rangle \otimes \langle\langle 4a-1,\ b]]$$

Proof. We have $q_1 \circ q_2 = \begin{pmatrix} 2 & 1 \\ 1 & 2a \end{pmatrix} \otimes [1,b]$. Then we easily see that $a(q_1 \circ q_2) = 1$, and therefore $q_1 \circ q_2 \cong \langle c\rangle \otimes \langle\langle c_1, d_1]]$ with suitable $c_1, d_1 \in A$. Computing the Witt-invariant of $q_1 \circ q_2$ we get $w(q_1 \circ q_2) = [(1-4a, -b]] = [(-c_1, -d_1]]$, and therefore $(-c_1, -d_1] \cong (1-4a, -b]$ (see [Kn]). Now we compare the norm forms of these algebras, obtaining $\langle\langle c_1, d_1]] \cong \langle\langle 4a-1, b]]$. This proves the lemma.

(8.2) Remark. If A is a semi local ring with $4 = 0$ then the product $\circ$ in Wq(A) is trivial as one directly sees from the proposition (8.1). The formula in (8.1) is not so nice because we are not able to decide, whether $c = 1$ or not. But in the special case $a = b$ we have

(8.3) Lemma. For any quadratic space $[1,a]$ over A it holds that $[1,a] \circ [1,a] \cong 2 \times [1,a]$.

Proof. We first show the assertion under the assumption $|A/m| \geq 4$ for all $m \in \max(A)$. In this case we can find $b \in A$ with $1-2b$, $1-4b \in A^*$ and $[1,a] \cong [1,b]$. Now we have

$$(8.4) \quad \begin{pmatrix} 2 & 1 \\ 1 & 2b \end{pmatrix} \perp \langle -1\rangle \perp \langle 1\rangle \cong \langle 1, 1, 2b-1, (2b-1)(1-4b)\rangle$$

and hence for $q \cong [1,b]$

$$q \circ q \perp 2 \times \mathbb{H} \cong \langle 1,1,2b-1,(2b-1)(1-4b)\rangle \otimes q$$

$$\cong \langle 1,1,2b-1, -(2b-1)\rangle \otimes q$$

because $4b-1 \in \underline{D}(q)^*$. Therefore $q \circ q \perp 2 \times \mathbb{H} \cong 2 \times q \perp 2 \times \mathbb{H}$, which implies $q \circ q \cong 2 \times q$ (see (4.3), chap. III). Take now any semi local ring A and consider the cubic Frobenius extension $B = A[X] / (X^3 + 6X^2 - X + 1)$ with the usual trace map s. Then over B we have $q \circ q \cong 2 \times q$ and therefore $s_*(q \circ q) \cong 2 \times s_*(q)$, i.e. $q \circ q \perp 4 \times \mathbb{H} \cong 2 \times q \perp 4 \times \mathbb{H}$, proving again $q \circ q \cong 2 \times q$ over A.

(8.5) Corollary. Let $q = \langle\langle a_1,\ldots,a_n,b]]$ be a quadratic Pfister space over A. Then for any integer $m \geq 1$ it holds that $q^m \cong 2^{(n+1)(m-1)} \times q$.

Proof. We set $\varphi = \langle\langle a_1,\ldots,a_n\rangle\rangle$ and $q_o = [1,b]$, i.e. $q = \varphi \otimes q_o$. Hence $b_q = \varphi \otimes b_{q_o}$ and $q \circ q \cong \varphi \otimes \varphi \otimes (q_o \circ q_o)$. Since $\langle 1,a\rangle \otimes \langle 1,a\rangle \cong 2 \times \langle 1,a\rangle$ for all $a \in A^*$ and $q_o \circ q_o \cong 2 \times q_o$, we get $q \circ q \cong 2^{n+1} \times q$. Now the corollary follows by a simple induction argument.

(8.6) Lemma. Let $q = \varphi \otimes [1,b]$ be a quadratic space, where $\varphi = \langle a_1,\ldots,a_n\rangle$, $a_i \in A^*$. If [q] is nilpotent in Wq(A), then [q] is a torsion element.

Proof. If φ is a Pfister space, then the lemma follows immediately from (8.5). Otherwise we construct for any sequence $\varepsilon = (\varepsilon_1,\ldots,\varepsilon_n)$, $\varepsilon_i = \pm 1$, the bilinear Pfister space $\gamma(\varepsilon,\varphi) = \langle\langle \varepsilon_1 a_1,\ldots,\varepsilon_n a_n\rangle\rangle$. Then we know that $\gamma(\varepsilon,\varphi) \otimes q \sim \bar{\varepsilon} \times \gamma(\varepsilon,\varphi) \otimes [1,b]$ where $\bar{\varepsilon} = \varepsilon_1 +\ldots+ \varepsilon_n$. Since [q] is nilpotent, it follows that $\bar{\varepsilon} \times [\gamma(\varepsilon,\varphi) \otimes [1,b]]$ is nilpotent, too. Hence $\bar{\varepsilon}^r \times [\gamma(\varepsilon,\varphi) \otimes [1,b]]^r = 0$ for some $r \geq 1$. Using (8.5) we get $\bar{\varepsilon}^r\, 2^{(n+1)(r-1)} \times [\gamma(\varepsilon,\varphi) \otimes [1,b]] = 0$. If $\bar{\varepsilon} \neq 0$ it follows that $[\gamma(\varepsilon,\varphi) \otimes [1,b]]$ is a torsion element and therefore $[\gamma(\varepsilon,\varphi) \otimes q]$ is torsion element. If $\bar{\varepsilon} = 0$, we have $[\gamma(\varepsilon,\varphi) \otimes q] = 0$. Using again the well-known relation $2^n \times [q] = \sum_{\varepsilon}[\gamma(\varepsilon,\varphi) \otimes q]$, we see that [q] is a torsion element. This proves the lemma.

(8.7) Theorem. For any semi local ring A it holds that $\mathrm{Nil}(Wq(A)) \subseteq Wq(A)_o \cap Wq(A)_t$.

Proof. Let $[q]$ be a nilpotent element with $[q]^m = 0$ for some $m \geq 1$. In particular, dim q must be even, i.e. $[q] \in W(A)_o$. Now we set

$$q \sim \psi_1 \otimes [1,b_1] \perp \dots \perp \psi_n \otimes [1, b_n]$$

with bilinear spaces ψ_i. We proceed by induction on n. The case $n = 1$ was treated in (8.6). Assume $n \geq 1$. Over $B = A(\wp^{-1}(-b_n))$ we get $q \otimes B \sim \psi_1 \otimes [1,b_1] \perp \dots \perp \psi_{n-1} \otimes [1,b_{n-1}]$. Hence $[q \otimes B] \in Nil(Wq(B))$ implies by induction that $[q \otimes B]$ is a torsion space. Hence $2^t \times [q \otimes B] = 0$ for some $t \geq 1$. Now (4.11) implies

$$2^t \times q \sim \varphi \otimes [1,b_n]$$

with some bilinear space φ. Hence $\varphi \otimes [1,b_n]$ is a nilpotent space, and therefore a torsion space (see (8.6)). This proves that q is a torsion space, and hence the theorem.

The converse of (8.7) is also true. The proof of this fact goes as in the field case. Namely, let $q = \langle a_1\rangle \otimes [1,b_1] \perp \dots \perp \langle a_n\rangle \otimes [1,b_n]$ be an even dimensional torsion space over A. We set $q_i = \langle a_i\rangle \otimes [1,b_i]$ for all $1 \leq i \leq n$, so that $q = q_1 \perp \dots \perp q_n$. Then for $N > n$ we get

$$q^N = \perp_{i_1+\dots+i_n=N} c_{i_1 \dots i_n} \times q_1^{i_1} \circ \dots \circ q_n^{i_n}$$

with $c_{i_n \dots i_n} \in \mathbb{N}$. Now, for every $1 \leq k \leq n$ we have $q_k^{i_k} \cong 2^{i_k-1} \times \langle d_k\rangle \otimes q_k$ with $d_k = a_k$ or 1. Hence every monomial $q_1^{i_1} \circ \dots \circ q_n^{i_n}$ is a multiple of $2^{\Sigma i_k - n} = 2^{N-n}$ and therefore $q^N \cong 2^{N-n} \times p$ for some quadratic space p. Hence $q^{N+1} \cong p \circ (2^{N-n} \times q)$, and since q is a torsion space, we detain $q^{N+1} \sim 0$ for a suitable $N > n$, i.e. q is a nilpotent space. Thus we have proved

(8.8) Theorem. $Nil\ (Wq(A)) = Wq(A)_o \cap Wq(A)_t$.

The corresponding result for W(A) follows now immediately, i.e.

(8.9) Theorem. $Nil\ (W(A)) = W(A)_o \cap W(A)_t$.

Proof. The inclusion $\supseteq$ can be proved in the same way as in the quadratic case. Take now $[\varphi] \in \mathrm{Nil}\,(W(A))$. Hence $[\varphi \otimes E_8] \in \mathrm{Nil}\,(Wq(A)) \subseteq Wq(A)_t$ (see 8.8)) , thus there is $s \geq 1$ such that $2^s \times [\varphi \otimes E_8] = 0$. In consequence we have $2^s \times \beta\,([\varphi \otimes E_8]) = 2^{s+3} \times [\varphi] = 0$, i.e. $[\varphi]$ is a torsion element. This proves the theorem.

(8.10) Remark. Let us consider any commutative ring A with 1. Let W(A) be the Witt ring of bilinear spaces over A. Then it is well-known that if $x \in W(A)$ is a torsion element, it follows that $2x$ is nilpotent (see [Dr], §2,Cor.5). Let now $L_1,\dots,L_n$ be bilinear spaces of rank 1 over A and define $\varphi = \langle 1,L_1\rangle \otimes \dots \otimes \langle 1,L_n\rangle = \langle\langle L_1,\dots,L_n\rangle\rangle$. Since $L_i \otimes L_i \cong \langle 1\rangle$, we get $\varphi^2 \cong 2 \times \varphi$ and in general $\varphi^N \cong 2^{N-1} \times \varphi$ for all $N \in \mathbb{N}$. Let us assume that the Pfister space φ is torsion. Using the cited result of Dress above, we see that $2 \times [\varphi]$ is nilpotent and hence $2^N \times [\varphi]^N = 0$ for some $N \geq 1$. Therefore we get $2^{2N-1} \times [\varphi] = 0$, i.e. the order of $[\varphi]$ is a power of 2 (see (6.1)). Let us now consider a bilinear space ψ over A which is diagonalizable, that is it has the form

$$\psi = L_1 \perp \dots \perp L_n$$

with suitable bilinear spaces L_i of rank 1 over A. Then we have (see (6.2)):

(8.11) Theorem. If $[\psi] \in W(A)$ is a torsion element, then the order of $[\psi]$ is a power of 2.

Proof. We argue as in the proof of (6.2). For any sequence of signs $\varepsilon = (\varepsilon_1,\dots,\varepsilon_n)$, $\varepsilon_i = \pm 1$, we define the n-fold Pfister space $\gamma(\varepsilon,\psi) = \langle\langle \varepsilon_1 L_1,\dots,\varepsilon_n L_n\rangle\rangle$. We have $\gamma(\varepsilon,\psi) \otimes \psi \sim \overline{\varepsilon} \times \gamma(\varepsilon,\psi)$ with $\overline{\varepsilon} = \varepsilon_1 + \dots + \varepsilon_n$. Since the order of $\gamma(\varepsilon,\psi)$ (which is indeed torsion) is a power of 2 (by the remark above), we see that the order of $\gamma(\varepsilon,\psi) \otimes \psi$ is a power of 2, too. Now our result follows from the well-known formula $2^n \times [\psi] = \sum_{\varepsilon} [\gamma(\varepsilon,\psi) \otimes \psi]$.

If $[q] \in Wq(A)$ is a torsion element, such that $\beta([q]) = [b_q]$ can be represented by a diagonalizable element, then we deduce from (6.8) and (8.11) that the order of $[q]$ is a power of 2.

It should be noted that in general $W(A)_t$ (and correspondingly $Wq(A)_t$) is not a 2-group (Karoubi). Namely, let M be a compact smooth manifold and $A= C^{\infty}(M,\mathbb{R})$ be the ring of smooth real functions on M. Let KO(M) be the Grothendieck ring of real vector bundles over M equiped with a non singular bilinear form. For any of such bundles ξ let $\Gamma(\xi)$ be the A-module of global smooth sections of ξ. Then $\Gamma(\xi)$ is a finitelly generated projective A-module equiped with a non singular bilinear form B, which is defined by $B(s,t)(x) = B_x(s(x), t(x))$ for all $x \in M$, $s,t \in \Gamma(\xi)$. Now, the map $[\xi] \longrightarrow [\Gamma(\xi), B]$ defines an isomorphism $KO(M) \xrightarrow{\sim} W(A)$ (see [M-H]). It is well-known that the group KO(M) may have torsion elements, whose orders are $\neq 2$. In particular, if $K = Quot(A)$ is the quotient field of A, we see that the induced homomorphism $W(A) \longrightarrow W(K)$ cannot be one to one. It would be interesting to classify the rings A, such that W(A) has only 2-torsion.

§ 9. An explicite description of $Wq(A)_t$.

Our aim is to determine explicitely the elements of $Wq(A)_t$ for any semi local ring A (see $[Pf]_3$). To do this we consider the problem in a more general setting. Let φ be a bilinear space over A. We denote the W(A)-submodule of Wq(A) consisting of all $[q] \in Wq(A)_0$ with $\varphi \otimes q \sim 0$ by $Ann(\varphi)$. Except for the trivial case $2 \in A^*$, $\varphi \sim 0$, which we shall exclude, $Ann(\varphi)$ is the full annullator of $[\varphi]$ in Wq(A), that is $Ann(\varphi) = \{[q] \in Wq(A) \mid [\varphi][q] = 0\}$. Using (6.3) we get

(9.1) $$Wq(A)_t = \bigcup_{n \geq 1} Ann(2^n \times \langle 1 \rangle) .$$

Since $Wq(A)_t = Wq(A)$ if $\langle 1 \rangle$ is a torsion space (i.e. A is non real), we shall assume that $2^n \times \langle 1 \rangle \not\sim 0$ for all $n \geq 1$. Thus through (9.1) we are led to calculate $Ann(2^n \times \langle 1 \rangle)$ for all $n \geq 1$. More generally, if φ is a bilinear Pfister space over A, we want to compute $Ann(\varphi)$ (see [B-K] for more details). Let us begin with the case $\varphi = \langle 1,c \rangle$, $c \in A^*$.

(9.2) Lemma. Let $\varphi = \langle 1,c \rangle$ be a binary Pfister space and (E,q) be a quadratic space over A. Then E has a decomposition

$$E = E_1 \perp \ldots \perp E_r \perp G$$

with two-dimensional quadratic spaces E_i $(1 \leq i \leq r)$ such that $\varphi \otimes E_i \sim 0$ and a space G, such that $\varphi \otimes G$ is anisotropic or $\dim G = 1$ (if $\varphi \otimes E$ is anisotropic, then $E = G$).
This result follows immediately (by induction) from the following

(9.3) Lemma. Assume $\dim E \geq 2$ and $\varphi \otimes E$ isotropic. Let $\{g_1, g_2\}$ be a basis of φ with $\varphi(g_1) = 1$, $\varphi(g_2) = c$, $\varphi(g_1, g_2) = 0$. Then for any primitive isotropic element $z \in \varphi \otimes E$ there exists $\sigma \in O^+(\varphi \otimes E)$, such that $\sigma(z) = g_1 \otimes x + g_2 \otimes y$ with $x, y \in E$ and $\langle x, y \rangle$ a non singular subspace of E.

Proof. Using the same reduction step as in (5.1), (5.2), chap.III, we may assume that A is a field. Since $\dim(\varphi \otimes E) \geq 4$, it is sufficient to show that there exist $x, y \in E$, such that $\langle x, y \rangle$ is a non singular subspace of E and $g_1 \otimes x + g_2 \otimes y$ is isotropic in $\varphi \otimes E$ (see (4.7)). If E is isotropic, we choose $x, y \in E$ such that $q(x) = q(y) = 0$, $(x, y) = 1$. If φ is isotropic, i.e. $c = -1$, we take any binary non singular subspace $F = \langle e, f \rangle$ of E with a canonical basis $\{e, f\}$, such that $q(e) \neq 0$. Then $x = f$, $y = \sigma_e(f)$ have the required property. Thus we may assume that both E and φ are non isotropic. From our hypothesis it follows that there is an isotropic vector $g_1 \otimes u + g_2 \otimes v \neq 0$ in $\varphi \otimes E$, i.e. $q(u) + cq(v) = 0$ with $u, v \in E$ not both zero. Since E is anisotropic, it follows that $q(u), q(v) \neq 0$. We set $a = q(u)$, $b = (u, v)$. Then $q(u) = -ca$ and the determinant of $\langle u, v \rangle$ is $-(b^2 + 4ca^2)$. In case $b^2 + 4ca^2 \neq 0$, we just take $x = u$, $y = v$. Let us now assume $b^2 + 4ca^2 = 0$. Since $-c$ is not a square, we must have $2 = 0$ and therefore $b = 0$. This implies $\dim E \geq 4$. Moreover E must contain a subspace of the form $[-ca, d'] \perp [a, d]$ with suitable $d, d' \in A$. Let $r, s \in E$ be vectors with $q(r) = d'$, $q(s) = d$, $(u, r) = (v, s) = 1$, $(r, s) = (u, s) = (r, v) = 0$. Since $r + s \neq 0$, this vector is not isotropic, and hence we can define $x = u$, $y = \sigma_{r+s}(v)$. It is easy to verify that these vectors have the required property, proving the lemma and consequently (9.2), too.

The lemma (9.2) was first proved by Elman and Lam in $[E\text{-}L]_1$ for fiels of characteristic $\neq 2$. According to (9.2) we are now led to consider binary spaces $[1, b]$ such that $\langle 1, c \rangle \otimes [1, b] \sim 0$. This means $[1, b] \cong \langle -c \rangle \otimes [1, b]$ (see (4.3), chap.III), that is $-c \in \underline{D}([1, b])^*$. Therefore there exist $a' \in A$, $d' \in A^*$ with $-c = a'^2 + a'd' + bd'^2$. Setting $a = a'd'^{-1}$, $d = d'^{-1}$ we get $b = -a - a^2 - d^2c$, i.e. $[1, b] = [1, -a-a^2-d^2c]$. Conversely, every space $[1, -a-a^2-d^2c]$ has the property $\langle 1, c \rangle \otimes$

$[1,-a-a^2-d^2c] \sim 0$. Thus we have proved the following

(9.4) Theorem. Let (E,q) be a quadratic space over A, such that $\varphi \otimes E \sim 0$. Then

$$E = \perp_{i=1}^{n} \langle c_i \rangle \otimes [1,-a_i-a_i^2-d_i^2c]$$

with suitable $c_i \in A^*$, $a_i, d_i \in A$.

(9.5) Corollary. Let $E = \langle\langle a_1,\ldots,a_n,b]]$ be a quadratic Pfister space over A, such that $\langle 1,c\rangle \otimes E \sim 0$. Then there exist $c_1,\ldots,c_n \in A^*$ and $a,d \in A$, such that

$$E \cong \langle\langle c_1,\ldots,c_n,-a-a^2-d^2c]] \ .$$

Proof. According to (9.4) we have $E = \perp_{i=1}^{n} \langle c_i' \rangle \otimes [1,-a_i-a_i^2-d_i^2c]$ with suitable $c_i', a_i, d_i \in A$. Since $c_1' \in \underline{D}(E)^* = N(E)$, we have $\langle c_1' \rangle \otimes E \cong E$, so that we may assume $c_1' = 1$. With $a = a_1$, $d = d_1$ we get $[1,-a-a^2-d^2c] \subset E$. Now our assertion follows from (4.1), chap. IV.

(9.6) Remark. Take $d \in \underline{D}([1,b])^*$. Then there exist $a,c \in A$ with $d = a^2 + ac + c^2b$. If $|A/m| \geq 7$ for all $m \in \max(A)$, then we can choose a,c with the property c, $c + 2a$, $c + 2a - 2 \in A^*$. Writing $\bar{a}$ for ac^{-1} we get $b = -\bar{a} - \bar{a}^2 - c^{-2}d$. Now let $\{e,f\}$ be a canonical basis of $[1,b]$ with $q(e) = 1$, $q(f) = b$, $(e,f) = 1$. Since $1 + 2\bar{a} = (c + 2a)c^{-1} \in A^*$, we obtain $\langle e,f\rangle = \langle e,f'\rangle$ where $f' = (1 + 2\bar{a})^{-1}(ce + f)$. But $q(f') = -[c(1 + 2\bar{a})]^{-2}d$ and $(e,f') = 1$, so that $[1,b] \cong [1,-\bar{c}^2d]$, where $\bar{c} = c^{-1}(1 + 2\bar{a})^{-1}$. It should be noted that $1 - 2\bar{c} = 1 - 2/c+2a = (c + 2a - 2)/c+2a \in A^*$. We remenber that this fact was used during the proof of (5.7).

Now, we shall consider the general case, i.e. we want to compute $\mathrm{Ann}(\varphi)$ for a bilinear n-fold Pfister space $\varphi = \langle\langle a_1,\ldots,a_n\rangle\rangle$ over the semi local ring A. A first approach to this end is the following

(9.7) Theorem. $\mathrm{Ann}(\varphi)$ is the submodule of $Wq(A)$ generated by the spaces of the following two types:

type 1 : spaces $[1,b]$ with $\varphi \otimes [1,b] \sim 0$

type 2 : spaces $\langle 1,-a\rangle \otimes [1,b]$ with $a \in \underline{D}(\varphi \otimes [1,b])^*$.

Proof. According to (9.2) and (9.4) we may assume that $\dim \varphi > 2$. Let

$E = \perp_{i=1}^{r} \langle a_i \rangle \otimes [i,b_i]$ be a space with $\varphi \otimes E \sim 0$. If $r=1$, then E is similar to a space of type 1, thus we may assume $r>1$. Now the proof proceeds by induction on dim E. First we assume that r is even. Since $\varphi \otimes E$ is isotropic, there exist elements $x_i \in \varphi \otimes [1,b_i]$ with $q(x_i) \in A^*$ for all $1 \leq i \leq r$ such that

$$a_1 q(x_1) + \ldots + a_r q(x_r) = 0$$

(see (5.2), chap. III). Hence

$$E \sim \perp_{i=1}^{r} \langle a_i \rangle \otimes \langle 1,-q(x_i) \rangle \otimes [1,b_i] \perp \perp_{i=1}^{r} \langle a_i q(x_i) \rangle \otimes [1,b_i].$$

This implies $\varphi \otimes \perp_{i=1}^{r} \langle a_i q(x_i) \rangle \otimes [1,b_i] \sim 0$ (see the remark (9.8) below). But the space $\perp_{i=1}^{r} \langle a_i q(x_i) \rangle \otimes [1,b_i]$ is isotropic, so that

$$0 \sim \varphi \otimes \perp_{i=1}^{r} \langle a_i q(x_i) \rangle \otimes [1,b_i] \sim \varphi \otimes G$$

where $\dim G < \dim E$. Our assertion follows by induction. Let us now assume that r is odd. We set $E' = \perp_{i=1}^{r-1} \langle a_i \rangle \otimes [1,b_i]$ with $r-1 \geq 2$. From $\varphi \otimes E \sim 0$ we deduce

$$\varphi \otimes E' \sim \langle -a_r \rangle \otimes \varphi \otimes [1,b_r] \quad ,$$

thus $\varphi \otimes E'$ must be isotropic, because $\dim \varphi \otimes E' > 2 \dim \varphi$ (use (4.3), chap.III). Using (5.2), chap.III again, we obtain elements $x_i \in \varphi \otimes [1,b_i]$ with $q(x_i) \in A^*$, $1 \leq i \leq r-1$, such that

$$a_1 q(x_1) + \ldots + a_{r-1} q(x_{r-1}) = 0 .$$

Hence

$$E \sim \perp_{i=1}^{r-1} \langle a_i \rangle \otimes \langle 1,-q(x_i) \rangle \otimes [1,b_i] \perp F$$

where $F = \perp_{i=1}^{r-1} \langle a_i q(x_i) \rangle \otimes [1,b_i] \perp \langle a_r \rangle \otimes [1,b_r]$.

Now $\varphi \otimes F \sim 0$ (see (9.8) below) and F is isotropic, so that $F \sim G$

with $\dim G < \dim E$, $\varphi \otimes G \sim 0$. Our assertion follows again by induction. This proves the theorem.

(9.8) Remark. Of course the spaces of type 2 belong to $\mathrm{Ann}(\varphi)$. Namely, if $a \in \underline{D}(\varphi \otimes [1,b])^*$, then $\varphi \otimes \langle 1,-a\rangle \otimes [1,b] = \varphi \otimes [1,b] \perp \langle -a\rangle \otimes \varphi \otimes [1,b] \cong \varphi \otimes [1,b] \perp -\varphi \otimes [1,b]$ (see (2.4), chap.IV).

According to (9.7) we have now to find all spaces of type 1 and 2 with respect to a given n-fold Pfister space $\varphi = \langle\langle a_1,\dots,a_n\rangle\rangle$. For all unproved details we refer the reader to § 5 and 6 of [B-K]. Up to the end of this section we shall make the following assumption

$$(*) \quad |A/m| \geq 3 \quad \text{for all} \quad m \in \max(A) .$$

Let $[1,b]$ be of type 1, i.e. $\varphi \otimes [1,b] \sim 0$. Then there exist $x,y \in \varphi$ with $\varphi(y) \in A^*$ and $\varphi(x) + (x,y) + \varphi(y)b = 0$ (see [B-K]). Putting $z = \varphi(y)^{-1}y$ we get $b = -\varphi(x)\varphi(z) - (x,z)$. Now (2.11), chap.IV, implies that $\varphi(x)\varphi(z) = (x,z)^2 + \varphi(v)$ for some $v \in \varphi'$ (where $\varphi = \langle 1\rangle \perp \varphi'$). Hence $b = -c - c^2 - \varphi(v)$, where $c = (x,z)$. Therefore $[1,b] = [1,-c-c^2-\varphi(v)]$, $v \in \varphi'$. Conversely, every space of the form $[1,-c-c^2-\varphi(v)]$, $v \in \varphi'$ is of type 1, because $\varphi \otimes [1,-c-c^2-\varphi(v)]$ is isotropic, and hence hyperbolic (see (3.2), chap.IV). Under the assumption $(*)$ in case $\dim \varphi > 2$, and under the assumption $|A/m| \geq 4$ for all $m \in \max(A)$ in case $\dim \varphi = 2$, it can be shown that we may always take $c = 0$ in the description above of the spaces of type 1 for φ (see (5.2) and (5.3) in [B-K]). Our next aim is to describe all spaces of type 2 for φ. We shall show that these spaces are equivalent to ortogonal sums of spaces of type 1. Let $E = \langle\langle -a,b]]$, $a \in \underline{D}(\varphi \otimes [1,b])^*$, be a space of type 2 for φ. Using lemma (4.3) in [B-K] we see that there exist $x,y \in \varphi$ with $\varphi(y) \in A^*$ and $a = \varphi(x) + (x,y) + \varphi(y)b$. We set $z = \varphi(y)^{-1}y$ and $\bar{a} = b + (x,z) + \varphi(x)\varphi(z)$. Hence $a = \varphi(y)\bar{a}$. Using (2.11), chap.IV again, we get $\varphi(x)\varphi(z) = (x,z)^2 + \varphi(u)$ for some $u \in \varphi'$, so that $\bar{a} = b + c + c^2 + \varphi(u)$, where $c = (x,z)$. From the well-known relation

$$(9.9) \qquad \langle 1,-\varphi(y)\bar{a}\rangle \sim \langle 1,-\varphi(y)\rangle \perp \langle \varphi(y)\rangle \otimes \langle 1,-\bar{a}\rangle$$

we obtain

$$E \sim \langle 1,-\varphi(y)\rangle \otimes [1,b] \perp \langle \varphi(y)\rangle \otimes \langle 1,-\bar{a}\rangle \otimes [1,b] .$$

We set $E_1 = \langle 1,-\varphi(y)\rangle \otimes [1,b]$ and $E_2 = \langle 1,-\bar{a}\rangle \otimes [1,b]$. Let $d \in A$, $v \in \varphi'$ be such that $\varphi(y) = d^2 + \varphi(v)$. Using the same line of reasoning as in the proof of (3.4), chap.IV and the relation (9.9), we easily see that it may be assumed $\varphi(v) \in A^*$. Since $\langle 1,\varphi(v)\rangle \otimes E_1$ is isotropic, it follows $\langle 1,\varphi(v)\rangle \otimes E_1 \sim 0$. Hence $E_1 \cong \langle a_1\rangle \otimes [1,b_1] \perp \langle a_2\rangle \otimes [1,b_2]$ with $\langle 1,\varphi(v)\rangle \otimes [1,b_i] \sim 0$, $i = 1,2$. But $\langle 1,\varphi(v)\rangle \subset \varphi$ implies $\varphi \otimes [1,b_i] \sim 0$, $i = 1,2$, too, so that our claim is true for E_1. Now we consider E_2. We have $\varphi(u) = \varphi(y)^{-2}[\varphi(x)\varphi(y) - (x,y)^2]$. An straightforward computation shows that we may assume $\varphi(x)\varphi(y) - (x,y)^2 \in A^*$ (see (6.2), [B-K]), i.e. $\varphi(u) \in A^*$. Since E_2 is round and represents the element $c + c^2 + b - \bar{a} = -\varphi(u)$, we get $E_2 \cong \langle -\varphi(u)\rangle \otimes E_2$, that is $\langle 1,\varphi(u)\rangle \otimes E_2 \sim 0$. From (9.2) we get $E_2 \cong \langle c_1\rangle \otimes [1,d_1] \perp \langle c_2\rangle \otimes [1,d_2]$ with $\langle 1,\varphi(u)\rangle \otimes [1,d_i] \sim 0$, $i = 1,2$. Since $\langle 1,\varphi(u)\rangle \subset \varphi$, we conclude that $\varphi \otimes [1,d_i] \sim 0$, $i = 1,2$. Thus the claim is true for E_2 and therefore for E. Summing up these results, we get (see (6.1), [B-K])

(9.10) Theorem. Let A be a semi local ring with the property $(*)$. Then for every even dimensional quadratic space E with $\varphi \otimes E \sim 0$ it holds

$$E \sim \langle a_1\rangle \otimes [1,b_1] \perp \ldots \perp \langle a_r\rangle \otimes [1,b_r] ,$$

where $b_i = -(c_i + c_i^2 + \varphi(v_i))$ with suitable $c_i \in A$, $v_i \in \varphi'$. If either $\dim \varphi > 2$ or $|A/m| \geq 4$ for all $m \in \max(A)$, we can even take $b_i = -\varphi(v_i)$ with $v_i \in \varphi'$, $1 \leq i \leq r$.

In particular we get

(9.11) Corollary. Let A be as in (9.10). Assume A to be real in case $2 \in A^*$. Then

$$Wq(A)_t = \sum_a W(A)\,[1,-a] ,$$

where a ranges over all sums of squares of A with $1 + 4a \in A^*$. If A is non real, then $Wq(A)_t = Wq(A)$.

(9.12) Corollary. Let A be a semi local ring as in (9.10). Then Wq(A) is torsion free if and only if every sum of squares a in A with $1 + 4a \in A^*$ has the form $b + b^2$ for some $b \in A$, and in the case $2 \in A^*$ the element -1 is not a square.

Proof. This follows from (9.11) and the fact that the space $[1,-a]$ is hyperbolic if and only if $a = b + b^2$ for some $b \in A$.
Now we make an application of these results. For any integer $n \geq 0$ one may ask, whether $I^n Wq(A)_0 \cap Wq(A)_t$ is generated (over W(A)) by torsion (n+1)-fold Pfister spaces. For any n this problem is open, even in the field case. The case $n = 0$ has been treated in (9.11). For $n = 1$ we have

(9.13) Proposition. $IWq(A)_0 \cap Wq(A)_t$ is generated by the torsion 2-fold Pfister spaces over A.

Proof. Let J be the W(A)-submodule of Wq(A), which is generated by the torsion 2-fold Pfister spaces. Of course $J \subseteq IWq(A)_0 \cap Wq(A)_t$. Take $[q] \in IWq(A)_0 \cap Wq(A)_t$. According to (9.11) we have $q \sim \langle a_1\rangle \otimes [1,b_1] \perp \ldots \perp \langle a_r\rangle \otimes [1,b_r]$ with torsion spaces $[1,b_i]$, $1 \leq i \leq r$. Now we argue by induction on r. If $r = 1$, then $[1,b_1] \sim 0$, because the Arf-invariant of $[1,b_1]$ must be trivial). Thus $q \sim 0$. Assume $r > 1$. For every $1 \leq i \leq r$ we have $\langle\langle a_i, b_i]] \in J$ and hence

$$q \sim [1,b_1] \perp -[1,b_2] \perp \mathop{\perp}_{i=3}^{r} \langle a_i\rangle \otimes [1,b_i] \quad (\text{mod } J)$$

Now $[1,b_1] \perp -[1,b_2] \cong \mathbb{H} \perp \langle c\rangle \otimes [1,b]$ with a suitable torsion space $[1,b]$. Therefore $q \sim \langle c\rangle \otimes [1,b] \perp \langle a_3\rangle \otimes [1,b_3] \perp \ldots \perp \langle a_r\rangle \otimes [1,b_r]$ (mod J). Since the right side of this relation lies in $IWq(A)_0 \cap Wq(A)_t$ and it has the length $r - 1$, we can apply induction to conclude that it is in J. This proves $[q] \in J$.

(9.14) Remark. Let $\varphi = \langle\langle a_1,\ldots,a_n\rangle\rangle$ be a bilinear Pfister space over A. It is not known, if the annullator of $[\varphi] \in W(A)$ in W(A) can be described in a similar way as $\mathrm{Ann}(\varphi)$. Even it is not known, if it is generated by binary spaces over A.

§ 10 On the classification of quadratic spaces.

In this section we shall throughout assume that A is a semi local ring with the property $(*)$ of § 9, i.e. $|A/m| \geq 3$ for all $m \in \max(A)$. As usual, $I_A \subset W(A)$ is the maximal ideal of even dimensional bilinear spaces over A. In this section we shall study the homomorphism

$$(10.1) \qquad c_A : Wq(A)_o \,/\, I_A^2\, Wq(A)_o \longrightarrow \Delta(A) \times Br(A)$$

given by

$$c_A([q]) = (a(q),\, w(q))$$

for all $[q] \in Wq(A)_o$. Whether c_A is injective or not, is a well-known outstanding problem, proposed by Pfister in $[Pf]_3$. As remarked before, if A is a field of characteristic 2, then c_A is actually one to one (see [Sa]). In this section we shall prove that $I_A^2\, Wq(A)_o = 0$ implies the injectivity of c_A (see $[E\text{-}L]_4$ for the field case and [Ma] for the case $2 \in A^*$). More generally, let us consider the total signature homomorphism.

$$(10.2) \qquad s : Wq(A) \longrightarrow \prod_{\bar\sigma \in \overline{Sig}(A)} \mathbb{Z}_{\bar\sigma}$$

($\mathbb{Z}_{\bar\sigma} = \mathbb{Z}$) given by $s([q]) = (\bar\sigma(q)) \in \prod_{\bar\sigma} \mathbb{Z}_{\bar\sigma}$. In (7.16) we have proved $Wq(A)_t = \mathrm{Ker}(s)$. The main result of this section is the following (compare $[E\text{-}L]_4$).

(10.3) Theorem. Let $I_A^2\, Wq(A)_o$ be torsion free. Then the map

$$b_A : Wq(A) \longrightarrow Q(A) \times Br(A) \times \prod_{\bar\sigma \in \overline{Sig}(A)} \mathbb{Z}_{\bar\sigma}$$

given by

$$b_A(x) = (\hat{a}(x),\, w(x),\, s(x)) \quad , \; x \in Wq(A) ,$$

is one to one (see § 2 for the definition of $Q(A)$ and $\hat{a}$).

This theorem means that assuming $I_A^2 Wq(A)_o$ to be torsion free, then the quadratic spaces over A are classified by their dimension, Arf-invariant, Witt-invariant and total signature. Since we have shown that $\mathrm{Ker}(\hat{a}) = I_A\, Wq(A)_o$ (see (2.5)), we only need to consider the homomorphism

$$(10.4) \qquad I_A\, Wq(A)_o \longrightarrow Br(A) \times \prod_{\bar{\sigma} \in \overline{Sig}(A)} \mathbb{Z}_{\bar{\sigma}}$$

given by $x \longrightarrow (w(x), s(x))$. We shall show that (10.4) is one to one. To prove this fact we may assume that A is connected. First let us introduce some notation. We denote the set of all elements $b \in A$ of the form $b = d + d^2 + \Sigma\, c_i^{\,2}$ with $1 + 4b \in A^*$ by $\Sigma(A)$. Let $\Sigma'(A)$ be the subset of $\Sigma(A)$ consisting of all sums of squares $b = \Sigma c_i^{\,2}$ with $1 + 4b \in A^*$. The main step in the proof of the theorem is the following

(10.5) Lemma. For $b \in \Sigma(A)$ we define $B = A(\wp^{-1}(b))$. If $I_A^2\, Wq(A)_o$ is torsion free, then $I_B^2\, Wq(B)$ is torsion free, too.

Proof of the theorem (10.3). Let q be a quadratic space over A with even dimension, $a(q) = 1$, $w(q) = 1$ and $s(q) = 0$. We want to show $q \sim 0$. Let us assume without restriction $\dim q > 0$. Since $a(q) = 1$, it follows that $[q] \in I_A Wq(A)_o$. Also $s(q) = 0$ implies $[q] \in Wq(A)_t$. In particular (see (9.11)

$$q \sim \langle a_1\rangle \otimes [1,-b_1] \perp \ldots \perp \langle a_r\rangle \otimes [1,-b_r]$$

with $b_i \in \Sigma(A)$ for all $1 \leq i \leq r$. To prove $q \sim 0$ we proceed by induction on r. If $r \leq 2$, then $a(q) = 1$, $w(q) = 1$ imply $q \sim 0$. Assume $r > 2$. Putting $b = b_1$ and $B = A(\wp^{-1}(b))$, we get $q_B \sim \langle a_2\rangle \otimes [1,-b_2] \perp \ldots \perp \langle a_r\rangle \otimes [1,-b_r]$ over B. Since $I_B^2\, Wq(B)$ is torsion free (see (10.5)), it follows by induction that $q_B \sim 0$. Therefore $q \sim \varphi \otimes [1,-b]$ with a suitable bilinear space $\varphi = \langle c_1,\ldots,c_{2s}\rangle$, where $2s \leq r$ (see (4.10)). The fact that φ is even dimensional follows from $a(q) = 1$ and (4.12). On the other hand we have $\varphi \sim \langle c_1\rangle \otimes \langle\langle c_1c_2, c_1c_3\rangle\rangle \perp \varphi_1$ with $\varphi_1 = \langle -c_1c_2c_3,\ldots,c_{2s}\rangle$, $\dim \varphi_1 = 2s - 2$. Hence

$$q \sim \varphi \otimes [1,-b] \sim \langle c_1\rangle \otimes \langle\langle c_1c_2, c_1c_3, -b]] \perp \varphi_1 \otimes [1,-b].$$

Since $[1,-b]$ is a torsion space and $I_A^2\, Wq(A)_o$ is torsion free, we get $\langle\langle c_1c_2, c_1c_3, -b]] \sim 0$ and $q \sim \varphi_1 \otimes [1,-b]$. Applying induction to

$\varphi_1 \otimes [1,-b]$ we get $q \sim 0$. This proves the theorem.

The rest of this section is devoted to prove the lemma (10.5). The proof goes over several steps (see $[E\text{-}L]_4$).

(10.6) <u>Lemma</u>. For any $a \in A$ with $1 + 4a \in A^*$ consider $B = A(\wp^{-1}(a))$. Then the ideal I_B^2 is generated additively by the Pfister spaces $\langle\langle c,\alpha\rangle\rangle$ with $c \in A^*$, $\alpha \in B^*$. Correspondingly, $I_B Wq(B)_o$ is generated by the Pfister spaces $\langle\langle c,\beta]]$ and $\langle\langle \gamma,d]]$ with $c \in A^*$, $d \in A$, $1 - 4d \in A^*$, $\gamma \in B^*$, $\beta \in B$, $1 - 4\beta \in B^*$.

<u>Proof</u>. We first consider I_B^2. Of course I_B^2 is additively generated by the Pfister spaces $\langle\langle\alpha,\beta\rangle\rangle$ with $\alpha,\beta \in B^*$. We consider $\langle\langle\alpha,\beta\rangle\rangle$, where $\alpha = a_o + a_1 z$, $\beta = b_o + b_1 z$ with $a_i, b_i \in A$ and $z^2 = z + a$. Since we may replace α and β by $\alpha\gamma^2$ and $\beta\delta^2$ for some $\gamma,\delta \in B^*$, respectively, it follows that we can assume $a_1, b_1 \in A^*$. Therefore $\langle 1, a_o + a_1 z\rangle = \langle 1, a_1(a_o a_1^{-1} + z)\rangle \sim \langle 1,-a_1\rangle \perp \langle a_1\rangle \otimes \langle 1, a_o a_1^{-1} + z\rangle$ and $\langle 1, b_o + b_1 z\rangle \cong \langle 1, -b_1(-b_o b_1^{-1} - z)\rangle \sim \langle 1, b_1\rangle \perp \langle -b_1\rangle \otimes \langle 1, -b_o b_1^{-1} - z\rangle$. Therefore we can assume $\alpha = a_o + z$ and $\beta = b_o - z$. If $a_o + b_o \in A^*$, then $\langle\langle \alpha,\beta\rangle\rangle \cong \langle\langle a_o + b_o, \delta\rangle\rangle$ with a suitable $\delta \in B^*$, proving our claim. Let us now consider the general case. Let $\delta = x + yz$ be a general element of B with undetermined $x,y \in A$. Then

$$(a_o+z)\delta^2 = a_o(x^2+y^2a) + (2x+y)ya + ((a_o+1)(2x+y)y + x^2+y^2a)\,z$$

Now we require

$$\text{(i)}\quad (a_o+1)(2x+y)y + x^2 + y^2a \in A^*$$

$$\text{(ii)}\quad \frac{a_o(x^2+y^2a) + (2x+y)ya}{(a_o+1)(2x+y)y + x^2+y^2a} + b_o \in A^*$$

Using the chinese remainder theorem we easily deduce from the hypothesis (*) that there exist $x,y \in A$ such that $\delta = x+yz \in B^*$, (i) and (ii) are satisfied. Now we replace a_o+z by $(a_o+z)\delta^2$. The conditions (i), (ii) reduce the problem (using the argument above) to the case $a_o+b_o \in A^*$, which is obvious. This proves the first assertion. In the second case we argue in a similar way. Let $\langle\langle\alpha,\beta]]$ be a generator of $I_B Wq(B)_o$,

α, $1-4\beta \in B^*$, where $\alpha = a_o+a_1z$, $\beta = b_o+b_1z$ $(a_i, b_i \in A)$. Replacing (if necessary) α by $\alpha\gamma^2$ and β by $(1+2\delta)^{-2}(\delta+\delta^2+\beta)$ with suitable $\gamma \in B^*$, $\delta \in B$, we may assume again $a_1, b_1 \in A^*$. Therefore

$$<<\alpha,\beta]] = <<-a_1b_1^{-1}(a_o' - b_1z) , \beta]] \qquad (\text{here } a_o' = -a_oa_1^{-1}b_1)$$
$$\sim <<a_1b_1^{-1},\beta]] \perp <-a_1b_1^{-1}> \otimes <<a_o'-b_1z,\beta]] .$$

Thus we only must consider the case $\alpha = a_o - b_1z$, $\beta = b_o + b_1z$. Again by a similar argument as in the first part of the proof, we are lead to consider the case $1-4(a_o+b_o) \in A^*$. Hence

$$<<a_o-b_1z,\ b_o+b_1z]] \cong [1,\ b_o+b_1z] \perp <a_o-b_1z> \otimes [1,\ b_o+b_1z]$$
$$\cong [1,\ a_o+b_o] \perp E$$

where $\dim E = 2$. Comparing the Arf-invariants we obtain $E \cong <\gamma> \otimes [1,\ a_o+b_o]$ with $\gamma \in B^*$, that is $<<\alpha,\beta]] \cong <<\gamma, a_o+b_o]]$. This concludes the proof of the lemma.

(10.7) <u>Lemma.</u> Let $B = A(\wp^{-1}(a)) = A \oplus Az$ be as in (10.6) and $s:B \longrightarrow A$ be the usual trace map $s(1) = 0$, $s(z) = 1$. Then for all $n \geq 1$ we have

$$s_*[I_B^n Wq(B)_o] \subseteq I_A^n Wq(A)_o$$

<u>Proof.</u> We know that $I_B^n Wq(B)_o$ is additively generated by the (n+1)-fold Pfister spaces $q = <<\alpha_1,\ldots,\alpha_n,\beta]]$, where in virtue of (10.6) we may assume $\alpha_1,\ldots,\alpha_{n-1} \in A^*$ and either $\alpha_n \in A^*$ or $\beta \in A$. The first case implies $s_*(q) \cong <<\alpha_1,\ldots,\alpha_n>> \otimes s_*([1,\beta])$, and hence $s_*(q) \in I_A^n Wq(A)_o$. The second case implies $s_*(q) \cong <<\alpha_1,\ldots,\alpha_{n-1},\beta]] \otimes s_*(<1,\alpha_n>)$, and since $s_*(<1,\alpha_n>) \in I_A$, we get again $s_*(q) \in I_A^n Wq(A)_o$. This proves the lemma.

(10.8) <u>Lemma.</u> Consider $B = A(\wp^{-1}(b))$ with $b \in \Sigma(A)$. If every torsion quadratic 3-fold Pfister space over A is hyperbolic , then the same holds over B.

Proof. Let q be a torsion quadratic 3-fold Pfister space over B. According to the remark (10.9) below we may assume $2 \times q \sim 0$ and hence $q \cong \langle\langle\alpha,\beta,-\gamma]]$ with $\gamma = \lambda + \lambda^2 + \delta^2$ in B (see (9.5)). Using (10.6) we may assume $\alpha \in A^*$. Hence $s_*(q) \cong \langle\langle\alpha\rangle\rangle \otimes s_*(\langle\langle\beta,-\gamma]])$, where s is the usual trace map $s: B \longrightarrow A$. Now we have $s_*(\langle\langle\beta,-\gamma]])$ $\in I_A\, Wq(A)_0 \cap Wq(A)_t$ (see (10.7)) and since this group is generated by the torsion quadratic 2-fold Pfister spaces (see (9.14)), we see that $s_*(\langle\langle\beta,-\gamma]])$ is a sum (in Wq(A)) of torsion 2-fold Pfister spaces. Hence $s_*(q)$ is a sum (in Wq(A)) of torsion 3-fold Pfister spaces. Our hypothesis implies $s_*(q) \sim 0$. Hence from (5.3) and (4.14) it follows that $q \cong q_o \otimes B$, where q_o is a quadratic 3-fold Pfister space over A. On the other hand the kernel of $I_A^2\, Wq(A)_0 \longrightarrow I_B^2\, Wq(B)_0$ is $I_A^2\,[1,-b]$, which is generated by torsion 3-fold Pfister spaces over A, that is $I_A^2\,[1,-b] = 0$ by hypothesis (see (4.12)). Therefore both q and q_o are torsion 3-fold Pfister spaces. Since $q_o \sim 0$ by hypothesis, we get $q \sim 0$. This proves the lemma.

(10.9) Remark. Let A be a semi local ring, such that every quadratic n-fold Pfister space q with $2 \times q \sim 0$ is isotropic (and hence hyperbolic). We claim that every torsion quadratic m-fold Pfister space with $m \geq n$ is isotropic (and hence hyperbolic) (see $[E\text{-}L]_4$). Namely, let p be a torsion m-fold Pfister space with $m \geq n$. Let us assume $2^{s+1} \times p \sim 0$, but $2^s \times p \not\sim 0$ for some $s \geq 1$. Thus $2 \times 2^s \times p \sim 0$, which implies (see (9.5)) $2^s \times p \cong \langle\langle c_1,\dots,c_{m+s-1}, -b]]$ with some $b = c + c^2 + d^2$ in A. In particular $2 \times [1,-b] \sim 0$ implies $2 \times \langle\langle c_1,\dots,c_n,-b]] \sim 0$, and hence by hypothesis $\langle\langle c_1,\dots,c_n,-b]]$ ~ 0. Therefore $2^s \times p \sim 0$, a contradiction. This proves the claim.

(10.10) Lemma. The following assertions are equivalent:

(i) $I_A^2\, Wq(A)_0$ is torsion free

(ii) every torsion quadratic 3-fold Pfister space over A is hyperbolic.

Proof. Clearly (i) implies (ii). Let us now assume (ii). Take $[q] \in I_A^2\, Wq(A)_0 \cap Wq(A)_t$ with q anisotropic. Assuming $\dim q > 0$ we have $2^t \times q \sim 0$ but $2^{t-1} \times q \not\sim 0$ for some $t \geq 1$. Replacing q by $2^{t-1} \times q$, we may assume without restriction that $2 \times q \sim 0$. Hence $q \cong \perp_{i=1}^{r} \langle a_i\rangle \otimes [1,-b_i]$ with $b_i = c_i + c_i^2 + d_i^2$ in A for all $1 \leq i \leq r$. We set $b = b_1$, $B = A(\wp^{-1}(b))$. If $r = 1$, then obviously

$q \sim 0$, a contradiction. Now we use induction on r to prove $q \sim 0$. Over B we have $q \otimes B \sim \perp_{i=1}^{r} \langle a_i \rangle \otimes [1,-b_i]$, and since (ii) holds over B (see (10.8)), we conclude by induction that $q \otimes B \sim 0$. Hence $q \sim \varphi \otimes [1,-b]$ with $\varphi \in I_A^2$ (see (4.12)). But $I_A^2[1,-b]$ is generated by torsion 3-fold Pfister spaces, thus $I_A^2[1,-b] = 0$ by hypothesis, and in particular $\varphi \otimes [1,-b] \sim 0$. This proves $q \sim 0$, a contradiction. Hence $I_A^2\, Wq(A)_o \cap Wq(A)_t = 0$.

Proof of the lemma (10.5). The assertion in (10.5) follows immediately from (10.8) and (10.10). In particular, the proof of (10.4) is complete.

Finally, let us consider a special case of theorem (10.4). To do this, we first prove the following

(10.11) Lemma. If A is a semi local ring, such that $I_A^n\, Wq(A) = 0$ for some $n \geq 2$, then -1 is a sum of 2^n squares in A.

Proof. If $2 \in A^*$, then $2^{n+1} \times \langle 1 \rangle \sim 0$ implies $2^n \times \langle 1 \rangle \cong 2^n \times \langle -1 \rangle$ (see (4.3), chap. III), and therefore -1 is a sum of 2^n squares in A. Now we treat the general case. We choose $h \in \mathbb{N}$ with $1-2h$, $1-4h \in A^*$ and define the quadratic Pfister space $2^n \times [1,h]$. The hypothesis implies $2^n \times [1,h] \sim 0$, that is $2^n \times [1,h] \cong 2^n \times \mathbb{H}$. In particular the associated bilinear spaces are isomorphic, that is

$$(10.12) \qquad 2^n \times \begin{pmatrix} 2 & 1 \\ 1 & 2h \end{pmatrix} \cong 2^n \times \begin{pmatrix} 0 & 1 \\ 1 & 0 \end{pmatrix}$$

Using the relation $\begin{pmatrix} 2 & 1 \\ 1 & 2h \end{pmatrix} \perp \langle -1 \rangle \cong \langle 1, 2h-1, (2h-1)(1-4h) \rangle$, we get from (10.12)

$$2^n \times \langle -1 \rangle \perp 2^n \times \begin{pmatrix} 0 & 1 \\ 1 & 0 \end{pmatrix} \cong 2^n \times \langle 1, 2h-1, (2h-1)(1-4h) \rangle \quad \text{and}$$

therefore, adding $2^n \times \langle -1 \rangle$ to both sides

$$2^{n+1} \times \langle 1 \rangle \perp 2^n \times \begin{pmatrix} 0 & 1 \\ 1 & 0 \end{pmatrix} \cong 2^n \times \langle 1,-1 \rangle \perp 2^n \times \langle 1-2h, (2h-1)(4h-1) \rangle$$

Now, $2h-1$, $4h-1 \in A^*$ are sums of 4 squares in $\mathbb{Z}$ and hence in A, and since $4 \leq 2^n$, they are norms of similitude of $2^n \times \langle 1 \rangle$ (see (2.19), chap. IV). Thus the last relation implies

$$\begin{aligned} 2^{n+1} \times \langle 1 \rangle \perp 2^n \times \begin{pmatrix} 0 & 1 \\ 1 & 0 \end{pmatrix} &\cong 2^n \times \langle 1,-1 \rangle \perp 2^n \times \langle 1,-1 \rangle \\ &\cong 2^n \times \langle 1,-1 \rangle \perp 2^n \times \begin{pmatrix} 0 & 1 \\ 1 & 0 \end{pmatrix} \end{aligned}$$

Using the cancellation theorem of Knebusch (see (4.5), chap. III), we get $2^n \times <1> \cong 2^n \times <-1>$. Thus -1 is a sum of 2^n squares in A.

Combining (10.11) and (10.4) we obtain (see [Ma] for the case $2 \in A^*$)

(10.13) Corollary. Let A be a semi local ring with $I_A^2 \, Wq(A)_o = 0$. Then c_A is one to one.

The assertion in (10.5) can be strengthened in the special case $I_A^n \, Wq(A)_o = 0$ in the following way.

(10.14) Proposition. Let A be a non real semi local ring. Take $b \in A$ with $1 + 4b \in A^*$ and set $B = A(\wp^{-1}(b))$. Then the condition $I_A^n \, Wq(A)_o = 0$ is equivalent with $I_B^n \, Wq(B)_o = 0$ for all $n \geq 1$.

Proof. First assume $I_A^n \, Wq(A)_o = 0$. For the usual trace map $s: B \longrightarrow A$ we obtain

$$s_*[I_B^n \, Wq(B)_o] \subseteq I_A^n \, Wq(A)_o = 0$$

Then if $q = <<\alpha_1,\ldots,\alpha_n,\beta]]$ is a $(n+1)$-fold Pfister space over B, we have $s_*(q) \sim 0$. Using (4.14) and (5.3) we deduce $q \cong q_o \otimes B$ with a $(n+1)$-fold Pfister space q_o over A, and since $q_o \sim 0$, we get $q \sim 0$. Thus $I_B^n \, Wq(B)_o = 0$. Let us now suppose $I_B^n \, Wq(B)_o = 0$. Let q be a m-fold quadratic Pfister space over A with $m \geq n+1$, which we assume to be anisotropic. Since A is non real, there exists $t \geq 1$ such that $<1,-(1+4b)>^t \otimes q \sim 0$ but $<1,-(1+4b)>^{t-1} \otimes q \not\sim 0$ (see (8.9)). Thus, replacing q by $<1,-(1+4b)>^{t-1} \otimes q$, we may assume that $<1,-(1+4b)> \otimes q \sim 0$, $q \not\sim 0$. Now (5.7) and (5.10) imply $q \sim <c> \otimes s_*(p)$ for a suitable m-fold Pfister space p over B, $c \in A^*$. But $p \sim 0$ by hypothesis, so that $q \sim 0$. This is a contradiction, proving $I_A^m \, Wq(A)_o = 0$ for all $m \geq n$

§ 11 The behaviour of Wq(A) by Galois extensions

Let A be a commutative ring and $i: A \longrightarrow B$ be an extension of rings. Let G be a finite group of automorphisms of B. We set $B^G = \{b \in B \mid g(b) = b$ for all $g \in G\}$. The extension $i: A \longrightarrow B$ is called a Galois extension with Galois group G, if the following conditions are satisfied (see [I-deM], [Kn-O]):

(i) $B^G = A$

(ii) for every $M \in \max(B)$ and every $g \in G$, $g \neq 1$, there exists

$b \in B$ such that $g(b) - b \notin M$.

Let B/A be a Galois extension with Galois group G. The trace $Tr_{B/A} : B \longrightarrow A$ is defined as follows : $Tr(b) = \sum_{g \in G} g(b)$ for all $b \in B$ (we omit the index B/A). The following facts are well-known:

i) A is a direct sumand of B, ii) there is an element $b \in B$ with $Tr(b) = 1$, iii) the symmetric bilinear form $Tr : B \times B \longrightarrow A$ given by $Tr(x,y) = Tr(xy)$ for all $x,y \in B$ is non singular, iv) B/A is a separable extension. In particular $i: A \longrightarrow B$ is a Frobenius extension. From now on we shall assume that A is a semi local ring. If $i: A \longrightarrow B$ is an extension of odd degree, then we know that the induced homomorphisms $i^*: Wq(A) \longrightarrow Wq(B)$, $i^*: W(A) \longrightarrow W(B)$ are one to one (see (6.9)). In this case we shall identify $Wq(A)$ and $W(A)$ with their images $i^*Wq(A)$ and $i^*W(A)$, respectively.

On the other hand G operates on $Wq(B)$ and $W(B)$ in the following way: every automorphism $g: B \xrightarrow{\sim} B$ over A induces an automorphism $g^*: Wq(B) \xrightarrow{\simeq} Wq(B)$ and $g^*: W(B) \xrightarrow{\simeq} W(B)$ (which is trivial over $i^*Wq(A)$ and $i^*W(A)$, respectively). Explicitely: let (E,q) be a quadratic space over B and $g \in G$. We define $g(E,q) = (E^g, q^g)$ taking E^g to be E as abelian group with a new B-module operation defined by $b \circ m = g^{-1}(b)m$, equipped with the quadratic form $q^g(m) = g[q(m)]$ for all $b \in B$, $m \in E$. For a bilinear space (E,b) we define (E^g, b^g) in a similar way. Let us write $Wq(B)^G$ and $W(B)^G$ for the subrings of $Wq(B)$ and $W(B)$, respectively, which are left fixed elementwise by G. If B/A has odd degree, then of course $Wq(A) \subseteq Wq(B)^G$ and $W(A) \subseteq W(B)^G$. We shall prove

(11.1) <u>Theorem</u>. Let B/A be a Galois extension of the semi local ring A of odd degree. Then $Wq(A) = Wq(B)^G$ and $W(A) = W(B)^G$.

The equality $W(A) = W(B)^G$ was first proved in [K-R-W]$_1$ (see [R-W]$_1$, [K-Sch] for the field case). For the proof of (11.1) we need the following formula:

(11.2) <u>Lemma</u>. Let (E,q) (or (E,b)) be a quadratic (or bilinear) space over B. Then

$$B \otimes Tr_*(E) \cong \perp_{g \in G} E^g$$

In particular for the composite $Wq(B) \xrightarrow{Tr_*} Wq(A) \xrightarrow{i^*} Wq(B)$

(or $W(B) \longrightarrow W(A) \longrightarrow W(B)$) we get $i^* \circ Tr_*(x) = \sum_{g \in G} g(x)$ for all $x \in Wq(B)$ (or $x \in W(B)$).

Proof. For example let us consider a quadratic space (E,q) over B. Then for the B-module E it holds $f : B \otimes E_A \xrightarrow{\sim} \bigoplus_{g \in G} E^g$, where $f(b \otimes m) = \bigoplus_{g \in G} g^{-1}(b)m$.
Now $B \otimes E_A$ carries the quadratic form $B \otimes Tr_*(q)$ and $\bigoplus_{g \in G} E^g$ carries the quadratic form $\perp_{g \in G} q^g$. Thus we have to check that f is actually an isomorphism of these quadratic spaces. This is a straightforward computation, so that we omit the details.

(11.3) Corollary. For every $x \in Wq(B)^G$ (or $x \in W(B)^G$) we have

$$i^*(Tr_*(x)) = [G : 1]x$$

This formula implies (for every Galois extension B/A)

$$(11.4) \qquad \begin{aligned} [G : 1]\, Wq(B)^G &\subseteq i^*\, Wq(A) \\ [G : 1]\, W(B)^G &\subseteq i^*\, W(A) \end{aligned}$$

In particular, $Wq(B)^G/i^*Wq(A)$ and $W(B)^G/i^*W(A)$ are abelian groups with [G:1]-torsion. It should be noted that during the proof of (10.2) and (10.3) we never used that A is semi local.

(11.5) Lemma. Let A be a semi local ring. Then every element $x \in Wq(A)$ (or $x \in W(A)$) satisfies a relation of the form $x^m + c_{m-1} x^{m-1} + \dots + c_r x^r = 0$ with $c_i \in \mathbb{Z}$ $(r \leq i \leq m)$. Of course one can choose m to be odd.

Proof. This is a well-known result for W(A) (see § 7). In § 8 we showed the relation $[1,b]^2 \cong 2 \times [1,b]$ for any binary quadratic space [1,b]. Hence for any space $q = \langle a\rangle \otimes [1,b]$ we have $q^3 \cong 2^2 \times q$. Since Wq(A) is additively generated by the spaces $\langle a\rangle \otimes [1,b]$ $(a, 1-4b \in A^*)$ if $2 \notin A^*$, and by the spaces $\langle a\rangle$ if $2 \in A^*$, our assertion follows at once.

(11.6) Lemma. Let p be an odd number > 1. If $x^m \in p\,Wq(A)$, resp. $x^m \in p\,W(A)$ for some $m \geq 1$, then $x \in p\,Wq(A)$, resp. $x \in p\,W(A)$, that is $Wq(A) / p\,Wq(A)$, resp. $W(A) / p\,W(A)$ has no nilpotent elements $\neq 0$.

Proof. We just consider the case $x \in Wq(A)$. Let us assume $x^m \in Wq(A)$ for some $m \geq 1$. We may assume that m is odd. We write $x = x_1+\dots+x_r$, where x_i is a class represented by a binary quadratic space ($1 \geq i \geq r$). The relation $x^m \in p\,Wq(A)$ implies $x_1^{mp}+\dots+x_r^{mp} \in pWq(A)$, and since $x_i^{mp} = 2^{mp-1}x_i$ for all i, we conclude that $2^{mp-1}x \in p\ Wq(A)$. Now let $s,t \in \mathbb{Z}$ be two integers with $1 = s\,2^{mp-1} + tp$. Then $x = s\,2^{mp-1}x + tpx \in p\,Wq(A)$. This proves our assertion.

Now we prove the theorem (11.1). Since the same arguments hold for Wq(A) and W(A), we shall only consider Wq(A).

Take $z \in Wq(B)^G$ and set $p = [G:1]$. According to (11.1) we have $pz \in Wq(A)$. Using the relation $z^n + c_{n-1} + \dots + c_1 z = 0$, $c_i \in \mathbb{Z}$ (see (11.5)), we get

$$(pz)^n + pc_{n-1}(pz)^{n-1} + \dots + p^{n-1}c_1(pz) = 0$$

and hence $(pz)^n \in p\,Wq(A)$. Now (11.6) implies $pz \in p\,Wq(A)$, i.e. there exist $x \in Wq(A)$ with $pz = px$. Since p is odd, the relation $p(z-x) = 0$ implies $z = x$, because Wq(A) has only 2-torsion. Thus the theorem (11.1) is proved.

(11.7) Remark. The arguments above prove also the following fact: let $i:A \longrightarrow C$ be an extension of semi local rings. Then $Wq(C) / i^*Wq(A)$ has no p-torsion for any odd number p. This can be proved as follows. Take $z \in Wq(C)$ with $pz \in i^*Wq(A)$, p odd. Using some relation $z^n + c_{n-1}z^{n-1} + \dots + c_1 z = 0$ in Wq(C) we conclude as above $(pz)^n \in pi^*Wq(A)$ and this implies (see the proof of (11.6)) $pz \in pi^*Wq(A)$, that is $z \in i^*Wq(A)$.

Appendix A

On the level of semi local rings

Let A be a semi local ring. We define the level of A as the number

(A.1) $$s(A) = \min \{r \mid -1 = a_1^2 + \ldots + a_r^2, a_i \in A\},$$

if -1 is a sum of squares in A, and $s(A) = \infty$ otherwise. If A is a field, it has been shown by Pfister (see $[Pf]_1$) that s(A) is always a power of 2 (or ∞). The purpose of this appendix is to prove similar results for s(A), where A is any semi local ring. Our main result says that s(A) has the form 2^n or $2^n - 1$ for any semi local ring with $s(A) < \infty$. The number $s(A) = 2^2 - 1 = 3$ actually occurs in the case $A = \mathbb{Z}/4\mathbb{Z}$, but it is not known, if any number $2^n - 1$ with $n > 2$ is the level of some semi local ring. The following result enables us particularly to treat the case $2 \in A^*$.

(A.2) Lemma. Let (E,q) be a round quadratic space over A. If the order of $[E] \in Wq(A)$ is 2^{t+1} $(t \geq 0)$, then $2^t \times E$ is the smallest multiple of E, which contains -E.

Proof. Assume $2^{t+1} \times E \sim 0$. Using the cancelation law (see (4.3), chap. III), we get $2^t \times E \cong -2^t \times E$, that is -E is contained in $2^t \times E$. Assume now $s \times E \cong -E \perp F$ for some $s \geq 1$. Hence $(s+1) \times E$ is isotropic, and choosing $r \geq 0$ with $2^{r+1} \geq s + 1 > 2^r$, we obtain $2^{r+1} \times E \sim 0$ (see (3.1), chap. IV). Thus $2^{t+1} \mid 2^{r+1}$, and in particular $s \geq 2^r \geq 2^t$. This proves the lemma.

Let us now consider a semi local ring A with $2 \in A^*$ and $s(A) < \infty$. Therefore $-1 = a_1^2 + \ldots + a_s^2$ with $a_i \in A$, $s = s(A)$. Then $(s+1) \times \langle 1\rangle$ is isotropic, and choosing $t \geq 0$ such that $2^{t+1} \geq s + 1 > 2^t$, we get $2^{t+1} \times \langle 1\rangle \sim 0$ (see (3.1), chap. IV). Thus $\langle 1\rangle$ is a torsion space, whose order in $Wq(A) = W(A)$ is 2^{t+1}, because if $2^t \times \langle 1\rangle \sim 0$, then $2^{t-1} \times \langle 1\rangle \cong -2^{t-1} \times \langle 1\rangle$, which implies that -1 is a sum of $2^{t-1} < s$ squares in A, a contradiction. Now using (A.2) we see that $2^t \times \langle 1\rangle$ is the smallest multiple of $\langle 1\rangle$, which contains $\langle -1\rangle$, i.e. $s = 2^t$. Thus we have proved

(A.3) Theorem. Let A be a semi local ring with $2 \in A^*$ and $s = s(A) < \infty$.

Then s is a power of 2 and <1> is a torsion element in W(A), whose order is 2s.

In the general case we only have the following information about the order of <1> in W(A).

(A.4) <u>Proposition.</u> Let A be a semi local ring with $s = s(A) < \infty$. Take $2^{t+1} \geq s + 1 > 2^t$. Then <1> is a torsion element in W(A) and $2^{t+2} \times$ $<1> \sim 0$.

<u>Proof.</u> From the definition of s and the choise of t, it follows that $2^{t+1} \times <1>$ is isotropic. Now our assertion follows directly from (3.5) chap. IV.

Let us further consider a semi local ring A with $s=s(A) < \infty$. We denote the order of <1> in W(A) by 2^r, and we choose t such that 2^{t+1} $s \geq 2^t$. Hence $2^{t+2} \geq 2^r$ (see (A.3)). On the other hand a result of Knebusch says that (see $[K]_4$).

(A.5) $$2^{r-2} < s \leq 2^r$$

Collecting the results above, we obtain $2^r \in \{ 2^{t+2}, 2^{t+1}, 2^t\}$. But according to (A.7) below, s has the form 2^n or $2^n - 1$ for some n, thus $s \in \{ 2^n, 2^n - 1, 2^{n-1}, 2^{n-1} - 1 \}$. Now we compare this two remarks, obtaining

(A.6) <u>Proposition.</u> i) The order of <1> is either s or 2s, provided that s is a power of 2.
ii) The order of <1> is $s + 1$ or $2(s + 1)$, provided that s has the form $2^n - 1$.

The rest of this appendix is devoted to the proof of the following result.

(A.7) <u>Theorem.</u> Let A be a semi local ring with $s(A) < \infty$. Then, either $s(A) = 2^n$ or $s(A) = 2^n - 1$ for some $n \geq 0$.

The proof of (A.7) includes several steps. In lemma (A.9) and (A.1) below we shall make the following assumption.

(A.8) $$|A/m| \geq 7 \quad \text{for all } m \in \max(A)$$

(A.9) Lemma. Assume (A.8). Let $u \in A^*$ be a unit, such that $u = a_1^2 + \ldots + a_s^2$, $a_i \in A$, $s \geq 4$. Then there exists a relation $u = b_1^2 + \ldots + b_s^2$ with $b_1, b_2 \in A^*$.

Proof. For any $d \in A$ with $1 + d^2 \in A^*$ we have

$$
\begin{aligned}
(1 + d^2)^2\, u = {} & (a_1 + da_2 + da_3 + d^2a_4)^2 & + \\
& (-a_2 + da_1 - da_4 + d^2a_3)^2 & + \\
& (a_3 + da_4 - da_1 - d^2a_2)^2 & + \qquad \text{(A.10)} \\
& (-a_4 + da_2 + da_3 - d^2a_1)^2 & + \\
& (a_5^2 + \ldots + a_s^2)(1 + d^2)^2 ,
\end{aligned}
$$

which is obviously equivalent with $u = a_1 2 + \ldots + a_s^2$ (the terms $a_5, \ldots, a_s$ appear only if $s > 4$). We first show that one can assume $a_1 \in A^*$. If $a_1 \notin A^*$, let m be a maximal ideal of A with $a_1 \in m$. Then there exists $2 \leq i \leq s$, such that $a_i \notin m$. We assume without restriction that $a_2 \notin m$. Using the chinese remainder theorem, we can choose $d \in A$ with $1 + d^2 \in A^*$ and such that

$$
\begin{aligned}
& d \notin m \quad , \quad a_2 + a_3 + da_4 \notin m \\
& d \in n \quad \text{for all} \quad n \neq m \quad , \quad n \in \max(A)
\end{aligned}
$$

To do this, we must assume $a_2 + a_3 \notin m$ or $a_4 \notin m$, which is no restriction, because if $a_2 + a_3 \in m$ and $a_4 \in m$, we can interchange a_3 and a_4 (note that we are assuming $a_2 \notin m$). We now use the relation (A.10) to get a new relation $u = b_1^2 + \ldots + b_s^2$ with $b_1 \notin m$. It should be noted that $b_1 \notin n$, whenever $a_1 \notin n$ for $n \in \max(A)$. Using this process, we arrive after finetly many steps to a relation $u = c_1^2 + \ldots c_s^2$ with $c_1 \in A^*$. Hence we may assume $a_2 \in A^*$. We use again the relation (1.10) (with some d) to define

$$
\begin{aligned}
b_1 &= (1 + d^2)^{-1}\, [a_1 + d(a_2 + a_3) + d^2a_4] \\
b_2 &= (1 + d^2)^{-1}\, [-a_4 + d(a_2 + a_3) - d^2a_1].
\end{aligned}
$$

Since $a_1 \in A^*$, we have $a_1 + (a_2 + a_3)X + a_4 X^2 \not\equiv 0 \pmod{m}$ and

$-a_4 + (a_2 + a_3)X - a_1X^2 \not\equiv 0 \pmod{m}$ for all $m \in \max(A)$. Now we can use (A.8) and the chinese remainder theorem to find $d \in A$, such that $1 + d^2 \in A^*$, $a_1 + d(a_2 + a_3) + d^2a_4$, $-a_4 + d(a_2 + a_3) - d^2a_1 \in A^*$. Using this d in the definition above, one easily sees that (A.10) gives the desired relation.

(A.11) Lemma. Assume (A.8). Let $c_m \in A$ be given for every $m \in \max(A)$ with $2 \notin m$. Then for every relation $u = a_1^2 + \ldots + a_s^2 \in A^*$ with $s \geq 4$ there exists a relation $u = b_1^2 + \ldots + b_s^2$ in A, such that

i) $b_1^2 - c_m \notin m$ for all $m \in \max(A)$ with $2 \notin m$

ii) $b_2 \notin m$ for all $m \in \max(A)$ with $2 \in m$.

Proof. In virtue of (A.9), we may assume that $a_1, a_2 \in A^*$. Consider $m \in \max(A)$ with $2 \notin m$ and set $c = c_m$. For any $d \in A$ with $1 + d \in A^*$, let $u = b_1^2 + \ldots b_s^2$ be the relation, given by (A.10), where b_1, b_2, b_3, b_4 are defined by the first four terms in (A.10), divided by $1 + d^2$, respectively, and $b_5 = a_5, \ldots, b_s = a_s$ if $s > 4$. We may assume that $a_2 + a_3 \notin m$, because if $a_2 + a_3 \in m$, then $a_2 + (-a_3) \equiv 2a_2 \not\equiv 0 \pmod{m}$, so that we can start with the relation $u = a_1^2 + (-a_3)^2 + \ldots a_s^2$. But this implies that $f(X) = [a_1 + (a_2 + a_3)X + a_4 X^2]^2 - (1 + X^2)^2 c \not\equiv 0 \pmod{m}$, since the coefficient of X is $2a_1(a_2 + a_3) \not\equiv 0 \pmod{m}$. Using (A.8) and the chinese remainder theorem, we can find $d \in A$, such that $1 + d^2 \in A^*$, $f(d) \notin m$ and $d \in n$ for all $n \in \max(A)$, $n \neq m$. Taking this d in the definition above, we get a relation $u = b_1^2 + \ldots b_s^2$ with $b_1^2 - c \notin m$ and $b_i \equiv \pm a_i \pmod{n}$ for all $n \neq m$, $1 \leq i \leq s$. Repeating this process for all maximal ideals m of A with $2 \notin m$, we obtain the desired relation.

The next lemma is the main step in the proof of theorem (A.7).

(A.12) Lemma. Let A be any semi local ring. If -1 is a sum of $2^{n+1} - 2$ squares in A with $n \geq 2$, then -1 is actually a sum of 2^n squares in A.

Proof. We set $\Psi = 2^n \times \langle 1 \rangle = \langle z_1, \ldots, z_{2^n} \rangle$ with $\Psi(z_i) = 1$, $(z_i, z_j) = 0$ for all $1 \leq i,j \leq 2^n$. We have $\Psi = \langle z_1 \rangle \perp \Psi'$, where $\Psi' =$

$< z_2, \ldots, z_{2^n} >$. Our hypothesis says $-1 = a_1^2 + \ldots a_s^2$, $a_i \in A$, where $s = 2^{n+1} - 2$. First we shall prove our assertion under the assumption (A.8). Therefore, using (A.11), we may change the relation $-1 = a_1^2 + \ldots + a_s^2$ in such a way that some a_i's fulfill some relations to be specified later. Now, multiplying the relation

(A.13) $$1 + a_1^2 + \ldots + a_s^2 = 0$$

with $1 + b^2$ for some $b \in A$, it follows that

(A.14) $$1 + b^2 + (a_1+ba_2)^2 + (ba_1-a_2)^2 + \ldots + (a_{s-1}+ba_s)^2 + (ba_{s-1}-a_s)^2 = 0$$

Note, that the left side of this relation is a sum of 2^{n+1} squares. Thus, defining the following elements from Ψ

$$x = bz_1 + (ba_{2^n-1} - a_{2^n})z_2 + (a_{2^n+1} + ba_{2^n+2})z_3 + \ldots + (a_{s-1} + ba_s)z_{2^n-1} + (ba_{s-1} - a_s)\, z_{2^n}$$

$$y = z_1 + (a_1 + ba_2)z_2 + (ba_1 - a_2)z_3 + \ldots + (a_{2^n-1} + ba_{2^n})\, z_{2^n}$$

we deduce from (A.14) that

(A.15) $$\Psi(x) + \Psi(y) = 0 .$$

Now we claim

(A.16) There exists a relation $-1 = a_1^2 + \ldots + a_s^2$ with the following property: one can find $b \in A$ and $z \in \langle y\rangle^\perp$ (see the notation above) such that
i) $\Psi(z) \in A^*$
ii) $\Psi(z)\,\Psi(x) - (z,x) \in A^*$.

Let us first show, how our assertion follows from the claim. Taking $b \in A$ and $z \in \langle y \rangle^\perp$ as in (A.16), we obtain from (A.15) $\Psi(z)\Psi(x) + \Psi(z)\Psi(y) = 0$. Now we use (2, 11), chap.IV, to find $x', y' \in \Psi'$, such that $\Psi(z)\Psi(x) = (z,x)^2 + \Psi(x')$ and $\Psi(z)\Psi(y) = (z,y)^2 + \Psi(y')$ (here we need (i)). Since $(z,y)=0$, it follows that

$$(z,x)^2 + \Psi(x') + \Psi(y') = 0$$

and setting $u = (z,x)\, z_1 + y' \in \Psi$, we obtain

$$\Psi(x') + \Psi(u) = 0.$$

Now (ii) implies $\Psi(x') \in A^*$, thus $-1 = \Psi(x')^{-1}\Psi(u)$, and since Ψ is round (see (2.19) ,chap.IV), we get finally $-1 = \Psi(w)$ with $w \in \Psi$. This means that -1 is a sum of 2^n squares in A.

Let us now sketch the proof of the claim (A.16) (see $[B]_5$ for more details). First we remark that an element $z \in \langle y \rangle^{\perp}$ has necessarily the form

$$z = -[b_2(a_1 + ba_2) + b_3(ba_1 - a_2) + \ldots + b_{2^n}(a_{2^n-1} + ba_{2^n})]z_1$$
$$+ b_2z_2 + \ldots + b_{2^n}z_{2^n}$$

with any $b_2, \ldots, b_{2^n} \in A$. Thus for a suitable relation $-1 = a_1^2 + \ldots + a_s^2$ we have to find $b, b_2, \ldots, b_{2^n} \in A$, such that (i) and (ii) are satisfied for x,y,z defined as above. Using the chinese remainder theorem one sees that it suffices to find over every residue classfield A/m, $m \in \max(A)$, elements $\bar{b}, \bar{b}_2, \ldots, \bar{b}_{2^n} \in A/m$, such that

i) $\Psi(\bar{z}) \neq 0$

ii) $\Psi(\bar{z})\Psi(\bar{x}) - (\bar{z},\bar{x})^2 \neq 0$.

Thus to prove (A.16) we may assume that A is a field with $|A| \geq 7$. The expressions $\Psi(z) = f(b, b_2, \ldots, b_{2^n})$ and $\Psi(z)\Psi(x) - (z,x)^2 = g(b, b_2, \ldots, b_{2^n})$ are two polynomials from $A[b, b_2, \ldots, b_{2^n}]$, such that f has degree 2 in every variable and g has degree 4 in b and 2 in the other variables $b_2, \ldots, b_{2^n}$. An easy computation shows that the polynomial f is always $\neq 0$ (to see this fact just compare the coefficients of b_2 and b_3). To ensure that $g \neq 0$, we change, if necessary, the relation $-1 = a_1^2 + \ldots + a_s^2$ in such a manner that

$a_1, a_2 \notin m$ for all $m \in \max(A)$ with $2 \in m$ and with $a_1{}^2 \not\equiv \alpha^{-1} a_{2^n}{}^{-1}$ (mod m) for all $m \in \max(A)$, where $2 \notin m$ and $\alpha = 2^n - 3 \notin m$ (see (A.11)). Under these conditions we see by a tedious computation that $f \neq 0$, $g \neq 0$ (see $[B]_5$ for details). Now we use $|A| \geq 7$ to find $b, b_2, \ldots, b_{2^n} \in A$, such that

$$f(b, b_2, \ldots, b_{2^n}) \neq 0$$

$$g(b, b_2, \ldots, b_{2^n}) \neq 0 .$$

Thus we have proved the claim and hence our lemma under the assumption (A.8). Let us now consider any semi local ring A. We construct the cubic Frobenius extension $B = A[X]/(X^3 + 6X^2 - X + 1)$ and its associated trace map s (see (2.3), chap. IV). Since $|B/M| \geq 7$ for all $M \in \max(B)$ and -1 is a sum of $2^{n+1} - 2$ squares in B, it follows that $B \otimes \Psi$ represents -1 over B. But $B \otimes \Psi$ is round, thus

$$B \otimes \Psi \cong - B \otimes \Psi$$

Taking the transfer of this relation, we get $s_*(B \otimes \Psi) \cong - s_*(B \otimes \Psi)$, that is

$$s_*(\langle 1\rangle) \otimes \Psi \cong - s_*(\langle 1\rangle) \otimes \Psi$$

We know that $s_*(\langle 1\rangle) \cong \langle 1\rangle \perp \begin{pmatrix} 0 & 1 \\ 1 & 6 \end{pmatrix}$ and $\begin{pmatrix} 0 & 1 \\ 1 & 6 \end{pmatrix} \cong - \begin{pmatrix} 0 & 1 \\ 1 & 6 \end{pmatrix}$. Therefore

$$\Psi \perp \begin{pmatrix} 0 & 1 \\ 1 & 6 \end{pmatrix} \otimes \Psi \cong - \Psi \perp \begin{pmatrix} 0 & 1 \\ 1 & 6 \end{pmatrix} \otimes \Psi .$$

Since $V\left(\begin{pmatrix} 0 & 1 \\ 1 & 6 \end{pmatrix} \otimes \Psi\right) \subseteq V(\Psi)$, we may apply the cancellation theorem (4.5), chap. III, to conclude that $\Psi \cong -\Psi$. This implies that Ψ represents -1, that is -1 is a sum of 2^n squares. This completes the proof of the lemma, and obviously, of the theorem (A.7).

(A.17) Remark. Let $A \longrightarrow B$ be any ring extension. Then $s(B) \leq s(A)$. Now we define for any integer $s \geq 1$ the ring

$$A_s = \mathbb{Z}[X_1, \ldots, X_s] / (1 + X_1{}^2 + \ldots + X_s{}^2)$$

Obviously $s(A_s) \leq s$. Let us assume that there exists a ring A with

$s(A) = s$. Take $-1 = a_1^2 + \ldots + a_s^2$, $a_i \in A$. Then we have a ring homomorphism $A_s \to A$, which is defined by $x_i \mapsto a_i$ (x_i = class of X_i in A_s). Hence $s = s(A) \leq s(A_s) \leq s$, that is $s(A_s) = s$. Thus, if we expect to find a ring of a given level s, we must show $s(A_s) = s$. For example, it is well-known that for any integer $n \geq 0$ there exists a field with level 2^n (see [Pf]$_1$). Hence $s(A_{2^n}) = 2^n$. We conjecture $s(A_s) = s$ for all integers $s \geq 1$. The following result support this conjecture.

(A.18) Proposition. For all $n \geq 0$ it holds

$$s(A_{2^n+1}) = 2^n + 1$$

Proof. Let us assume $s(A_{2^n+1}) \leq 2^n$. Then there exists $\bar{f}_1, \ldots, \bar{f}_{2^n} \in A_{2^n+1}$, such that

$$-1 = \bar{f}_1^{\,2} + \ldots + \bar{f}_{2^n}^{\,2} .$$

In virtue of the relation $-1 = x_1^2 + \ldots + x_{2^n+1}^2$ in A_{2^n+1} we can find representatives $f_i \in \mathbb{Z}[X_1, \ldots, X_{2^n+1}]$ of the $\bar{f}_i$'s , which which are linear in X_{2^n+1} . Hence we have $f_i = g_i + X_{2^n+1} h_i$ with $g_i, h_i \in \mathbb{Z}[X_1, \ldots, X_{2^n}]$.

The relation above implies in $\mathbb{Z}[X_1, \ldots, X_{2^n+1}]$ the following relation

$$1 + \sum_{i=1}^{2n} f_i^{\,2} = p(X)(1 + X_1^{\,2} + \ldots + X_{2^n+1}^2) .$$

Comparing the coefficients of X_{2^n+1} and $X_{2^n+1}^2$, we obtain

$$p(X) = h_1^{\,2} + \ldots + h_{2^n}^2 \quad \text{and}$$

$$1 + \sum_{i=1}^{2n} g_i^{\,2} = \left(\sum_{i=1}^{2n} h_i^2\right)\left(1 + X_1^{\,2} + \ldots + X_{2^n}^2\right)$$

(A.19) $$\sum_{i=1}^{2n} g_i h_i = 0$$

Now we write $g_i = g_{oi} + \cdot\cdot\cdot + g_{ri}$, $h_i = h_{oi} + \cdot\cdot\cdot + h_{ti}$, where g_{ki} , h_{ki} are the homogeneous parts of degree k of g_i and h_i respectively. Of course, we may assume that $g_{ri} \neq 0$ and $h_{tj} \neq 0$ for some i and j (note that $t=r-1$, $r \geq 1$). Now, comparing the terms of highest degree in the equalities (A.19) we get

$$\sum_{i=1}^{2^n} g_{ri}^2 = \left(\sum_{i=1}^{2^n} h_{ti}^2\right)\left(X_1^2 + \cdot\cdot\cdot + X_{2^n}^2\right)$$

(A.20) $$\sum_{i=1}^{2^n} g_{ri} h_{ti} = 0$$

and therefore

$$X_1^2 + \cdot\cdot\cdot + X_{2^n}^2 = \frac{\left(\sum_{i=1}^{2^n} g_{ri}^2\right)\left(\sum_{i=1}^{2^n} h_{ti}^2\right)}{\left(\sum_{i=1}^{2^n} h_{ti}^2\right)^2}$$

From (2.23), chap. IV and (A.20) we conclude that

$$X_1^2 + \ldots + X_{2^n}^2 = l_1^2 + \ldots + l_{2^n-1}^2$$

with $l_i \in Q(X_1,\ldots,X_{2^n})$. This is a contradiction to a well-known theorem of Cassels (see[Ca],[L]). Thus we have proved $s(A_{2^n+1}) = 2^n + 1$.

Supported by such a result and (A,7), one may expect that there are semi local rings with level 2^n-1 for any integer $n \geq 1$. For n= 2 we have $s(\mathbb{Z}/4\mathbb{Z}) = 2^2 - 1$ (but also $= 2^1 + 1$). The natural candidate for the level $2^n - 1$ is the local ring $B_n = (A_{2^n-1})_m$, where $m \subset A_{2^n-1}$ is the maximal ideal $(2, X_1 - 1,\ldots, X_{2^n-1} - 1)$. But we do not know, how to prove

$$s(B_n) = 2^n - 1.$$

(A.21) Remark. We know that for any semi local ring A the torsion subgroup $Wq(A)_t$ of $Wq(A)$ is a 2-group. Let $h(A) = 2^t$ be the smallest power of 2 with $2^t\, Wq(A)_t = 0$ ($t = \infty$ is allowed). We call h(A) the hight of A (see [K - Sh]). Let us assume $s = s(A) < \infty$. Then $s = 2^n$ or $s = 2^n - 1$ for some $n \geq 0$ (see (A.7)). Hence for any $a \in A$ with $1 - 4a \in A^*$, the quadratic space $2s \times [1,a]$ or $2(s+1) \times [1,a]$ is isotropic (the same fact holds for the bilinear spaces $2s \times \langle 1 \rangle$ and

$2(s+1) \times <1>$). Hence $2s \times [1,a] \sim 0$ if $s = 2^n$ and $2(s+1) \times [1,a] \sim 0$ if $s = 2^n - 1$ (in the bilinear case we have $2s \times <1> \sim 0$ if $2 \in A^*$, and in general $2^2 s \times <1> \sim 0$ or $2^2(s+1) \times <1> \sim 0$ if $s = 2^n$ or $s = 2^n - 1$, respectively). Thus

$$h(A) \mid 2s(A) \quad \text{or} \quad h(A) \mid 2(s(A)+1)$$

In the case $2 \in A^*$ we deduce from (A.3) that $h(A) = 2s(A)$. If $2 \notin A^*$, then the situation is more involved. For example, for $A = \mathbb{Z}/4\mathbb{Z}$ we have $h(\mathbb{Z}/4\mathbb{Z}) = 2$, but $s(\mathbb{Z}/4\mathbb{Z}) = 3$, that is $h \neq 2(s+1)$. Correspondingly, for $\mathbb{Z}/16\mathbb{Z}$ we have $s(\mathbb{Z}/16\mathbb{Z}) = 4$ and $h(\mathbb{Z}/16\mathbb{Z}) = 2$. It would be interesting to know exactly, which relation holds between h(A) and s(A) in the case $2 \notin A^*$.

The following proposition gives us some informations about h, when we extend the ring A to a quadratic separable algebra $A(\wp^{-1}(a))$, $1+4a \in A^*$.

(A.22) <u>Proposition.</u> Let $B = A(\wp^{-1}(a))$ be a quadratic separable extension of A. Then

$$h(B) \mid 2h(A)$$

<u>Proof.</u> Take $x \in Wq(B)_t$ and assume $h = h(A) < \infty$. Let $s: B \longrightarrow A$ be the usual trace map, given by $s(1) = 0$, $s(\delta) = 1$, where $B = A \oplus A\delta$, $\delta^2 = \delta + a$. Then $0 = hs_*(x) = s_*(hx)$. From (5.2), chap.V, we get $hx \in \mathrm{Im}(i^*)$, when $i: A \longrightarrow B$ is the natural inclusion. Hence $hx = i^*(y)$ for some $y \in Wq(A)$. Now we can consider the trace map $Tr_{B/A}: B \longrightarrow A$ (see § 11, chap. V). From (11.3), chap.V, we get

$$2hx = i^*[Tr_*(hx)] = i^*[hTr_*(x)] = 0,$$

that is $h(B) \mid 2h(A)$.

Appendix B

The u-invariant

Let A be a semi local ring. We define the u-invariant of A as the number

(B.1) $u(A) = \max \{\dim q \mid q$ anisotropic torsion space over A $\}$

(B.2) Examples. 1. If A is complete in the r-adic topology (r=Jacobson radical), then, according to (1.4), chap.V, we have $u(A) = \max\{u(A/m) \mid m \in \max(A)\}$. Thus, if $A = k[[X]]$ is a ring of power series over the field k, we have $u(A) = u(k)$.
2. For any integer $n \geq 0$ there exists a field with u-invariant 2^n, namely $k_n = \mathbb{C}((X_1)) \ldots ((X_n))$ (here $\mathbb{C}((X_1))$ is the field of formal power series in X_1 over the field $\mathbb{C}$ of complex numbers). Correspondingly, for $k = \mathbb{C}(X_1, \ldots, X_n, \ldots)$, we get $u(k) = \infty$.
3. Let us consider again $A = k[[X]]$ with one variable X. Let $K = k((X))$ be the quotient field. If $ch(k) \neq 2$, then $u(K) = 2u(k)$ by a well-known theorem of Springer. If $ch(k) = 2$, we do not have such a theorem. In any case, it holds that $u(K) \geq 2u(k)$, because if q is anisotropic of dimension $u(k)$ over k, then $q \perp \langle X \rangle \otimes q$ is anisotropic over K, too.

The definition (B.1) of the u-invariant was first introduced by Elman and Lam in $[E\text{-}L]_1$ for fields of characteristic $\neq 2$. If $2 \notin A^*$, then obviously $u(A)$ is even. The same holds for formally real semi local rings, because $Wq(A)_t \subseteq Wq(A)_o$. It should be noted that if A is a non real semi local ring with $2 \in A^*$, then any quadratic space q over A with $\dim q = u(A)$ is universal, i.e. it represents all units of A. To see this, we just apply (5.2), chap.III to the isotropic space $q \perp \langle -a \rangle$, where $a \in A^*$.
A long standing problem about the u-invariant says, whether $u(A)$ is a power of 2 or not. This problem is unsolved till now. An extensively treatment of this problem has been done by Elman and Lam in $[E\text{-}L]_1$ and $[E\text{-}L]_2$. Particularly they have shown that, if the quaternion algebras over a field of characteristic $\neq 2$ form a subgroup of the Brauer group of the field, then the u-invariant can only be 1,2,4 or 8. In this appendix we want to prove a similar result for semi local rings.

(B.3) Remarks. (1) Let $I \subset W(A)$ be the ideal of even dimensional bilinear spaces over A. Then, if A is non real and $u(A) < \infty$, it follows that $I^{r-1}Wq(A)_o = 0$ for any r with $2^r > u(A)$. To see this, just note that $I^{r-1}Wq(A)_o$ is generated by the quadratic Pfister spaces q with $\dim q = 2^r > u(A)$ (and apply (3.2), chap.IV). If A is real, we expect the following result: if $u(A) < \infty$ and $2^r > u(A)$, then $I^{r-1}Wq(A)_o$ is torsion free. This assertion can easily be shown in the field case, using the "Hauptsatz" of Arason and Pfister (see $[E\text{-}L]_1$). For a semi local ring we can prove directly only the following special case: if $u(A) < 8$, then $I^2Wq(A)_o$ is torsion free. This follows from the definition of u(A) and (10.10), chap.V.

(2) If $1 < u(A) < \infty$ and $I^2Wq(A)_o$ is torsion free, then u(A) is even. To show this, we may assume $2 \in A^*$, because otherwise u(A) is always even. Let us assume that u(A) is odd. Hence A is non real and we have $I = I_t = Wq(A)_o$. Consider q anisotropic with $\dim q = u(A) = u$. Hence q is universal. In particular $d(q) = (-1)^{u(u-1)/2} \det(q) \in \underline{D}(q)$. This implies

$$q \cong \langle d(q)\rangle \perp q_o \quad ,$$

where $\dim q_o = u-1$. Therefore $d(q_o) = 1$, because $d(q_o) = (-1)^{(u-1)^2}$ and $u-1$ is even. Hence $q_o \in I^2$, and for any $a \in A^*$ we have

$$\langle 1, -a\rangle \otimes q_o \in I^3 = I_t^3 = 0 \quad ,$$

that is $q_o \cong \langle a\rangle \otimes q_o$. In particular $1 \in \underline{D}(q_o)$ and hence $a \in \underline{D}(q_o)$ for all $a \in A^*$. This is a contradiction, because $-d(q) \in \underline{D}(q_o)$ implies that q is isotropic.

(3) Let us assume $u = u(A) > 2$ and $I^2Wq(A)_o$ torsion free. We assert that there exists $[q] \in IWq(A)_o \cap Wq(A)_t$, such that q is anisotropic and $\dim q = u$. To prove this, we take $[q] \in Wq(A)_t$ with q anisotropic and $\dim q = u$. We have $2^n \times q \sim 0$ for some $n \geq 1$. Using the lemma (B.4) below (see [B-K]), we get $2^n \times \nabla(q) \sim 0$, where $\nabla(q)$ denotes the discriminant form of q (see (3.20), chap.II). Hence

$$[q \perp \nabla(q)] \in Wq(A)_t \quad ,$$

and since $a(q \perp \nabla(q) = 1$, we obtain $[q \perp \nabla(q)] \in IWq(A)_o \cap Wq(A)_t$. Let us now assume that our assertion is false for A, i.e. all anisotropic spaces in $IWq(A)_o \cap Wq(A)_t$ have dimension $\leq u-2$ (according to (2) u must be even). This implies

$$q \perp \nabla(q) \cong 2 \times \mathbb{H} \perp q_o \quad ,$$

where $\dim q_o = u-2$. Using the cancellation theorem and $2 \times \mathbb{H} \cong \nabla(q) \perp -\nabla(q)$, we get $q \cong -\nabla(q) \perp q_o$. On the other hand, we have $[q_o] \in IWq(A)_o \cap Wq(A)_t$ and hence $[<1,-a> \otimes q_o] \in I^2Wq(A)_o \cap Wq(A)_t = 0$ for all $a \in A^*$. Therefore $q_o \cong <a> \otimes q_o$ for all $a \in A^*$, showing that q_o is universal. Hence $q \cong -\nabla(q) \perp q_o$ must be isotropic. Contradiction.

During the proof of (3) above we have used the following result

(B.4) <u>Lemma.</u> Let φ be a bilinear Pfister space over A and q be a quadratic space, such that $\varphi \otimes q \sim 0$. Then $\varphi \otimes \nabla(q) \sim 0$.

<u>Proof.</u> Of course, we can assume that φ is anisotropic. According to (9.7), chap.V we can write

$$q \sim <a_1> \otimes F_1 \perp \ldots \perp <a_s> \otimes F_s \perp <c_1> \otimes [1,b_1] \perp \ldots$$
$$\perp <c_r> \otimes [1,b_r]$$

where the F_i's are spaces of type 2 and the $[1,b_i]$'s are of type 1 . We get from this relation $a(q) = [A(\wp^{-1}(-b_1) \circ \ldots \circ A(\wp^{-1}(-b_r))]$. Since $\nabla(q)$ is the norm of the representative of $a(q)$, we deduce from this last relation, using the isomorphism

$$[1,b] \perp -[1,b'] \cong \mathbb{H} \perp [\nabla(A(\wp^{-1}(-b)) \circ A(\wp^{-1}(-b')))] \quad ,$$

that $\nabla(q) \sim <d_1> \otimes [1,b_1] \perp \ldots \perp <d_r> \otimes [1,b_r]$ with suitable $d_i \in A^*$. Hence $\varphi \otimes \nabla(q) \sim 0$.

(B.5) <u>Proposition.</u> For any semi local ring u cannot be 3, 5 or 7.

<u>Proof.</u> Of course we can assume that A is non real and $u(A) \leq 7$. Hence $I^2Wq(A)_o$ is torsion free (see remark (B.3)(1)), which implies $u(A) = 1$ or $u(A)$ is even (see (B.3)(2)). This proves the proposition.

The rest of this appendix is devoted to the study of semi local rings with $u \leq 4$. We begin by proving the following result (see $[E-L]_2$).

(B.6) <u>Proposition.</u> For any semi local ring the following assertions are equivalent.

i) $u(A) \leq 4$

ii) $I^2Wq(A)_o$ is torsion free and every $x \in IWq(A)_o \cap Wq(A)_t$ is represented by a 2-fold Pfister space.

Proof. (ii) $\Rightarrow$ (i). Let us assume without restriction that $u(A) > 2$, and take $0 \neq [q] \in IWq(A)_o \cap Wq(A)_t$. From (B.3)(3) follows that one can take q anisotropic with $\dim q = u$. Thus (ii) implies (i).
(i) $\Rightarrow$ (ii). Let us assume $u(A) \leq 4$, and take $[q] \in I^2Wq(A)_o \cap Wq(A)_t$. Let us assume $[q] \neq 0$, q anisotropic. Then $\dim q \leq 4$, and since $a(q) = 1$, we must have $\dim q = 4$ if $[q] \neq 0$. Hence $q \cong \langle c\rangle \otimes \langle\langle a,b]]$, and since $w(q) = 1$, we get $q \sim 0$. This is a contradiction, proving that $I^2Wq(A)_o$ is torsion free. Let us now take $[q] \in IWq(A)_o \cap Wq(A)_t$ and with $\dim q \leq 4$. If $\dim q = 2$, we conclude that $q \sim 0$, because $a(q) = 1$. Assuming that $\dim q = 4$ (and hence $u=4$), we deduce from $a(q) = 1$ that $q \cong \langle c\rangle \otimes \langle\langle a,b]]$. But $I^2Wq(A)_o$ torsion free implies $\langle 1,-c\rangle \otimes q \sim 0$, that is $q \cong \langle c\rangle \otimes q$, and hence $q \cong \langle\langle a,b]]$, proving (i).

Using this proposition we now can characterize the case $u \leq 4$ in the following way (see $[E\text{-}L]_2$).

(B.7) Theorem. Let A be a semi local ring. Then the following assertions are equivalent.

(i) $u(A) \leq 4$

(ii) $I^2Wq(A)_o$ is torsion free and the classes of quaternion algebras $(a,w]$ over A, where $a \in A^*$, $w = c + c^2 + d^2$ $(c,d \in A,\ 1 + 4w \in A^*)$ form a subgroup in $Br(A)$.

Proof. (i) $\Rightarrow$ (ii). Assume $u(A) \leq 4$. Hence $I^2Wq(A)_o$ is torsion free (see (B.6)). Now let us consider two quaternion algebras $Q_1 = (a,w]$, $Q_2 = (b,u]$ of the form described in (ii). We define $q_1 = \langle\langle -a,-w]]$, $q_2 = \langle\langle -b,-u]]$ and $q = q_1 \perp q_2$. Then it follows that $[q] \in IWq(A)_o \cap Wq(A)_t$ and $w(q) = [Q_1 \otimes Q_2]\, w(\delta(q_1) \otimes \nabla(q_2))$ (see 3.22), chap. II). But $\nabla(q_2) \cong \mathbb{H}$, so that $w(q) = [Q_1][Q_2]$. Using (B.6)(ii) we can write $q \sim \langle\langle x,y]]$ with suitable $x,y \in A$, and since $2[q] \in I^2Wq(A)_o \cap Wq(A)_t = 0$, we conclude that $2 \times \langle\langle x,y]] \sim 0$. Hence $\langle\langle x,y]] \cong$

$\langle\langle -e,-v]]$ with $e \in A^*$, $v = c + c^2 + d^2$, $1 + 4v \in A^*$ (see (9.5), chap.V). Thus we have proved $[Q_1][Q_2] = [(e,v]]$ in Br(A). This proves (ii).

(ii) ⇒ (i). It suffices to prove the assertion (ii) of (B.6). Thus let us consider $[q] \in IWq(A)_o \cap Wq(A)_t$ with q anisotropic. Hence for any $a \in A^*$ we have $\langle a\rangle \otimes q \cong q$, since $I^2Wq(A)_o$ is torsion free. In particular $A^* \subset \underline{D}(q)$. Using (9.4), chap.V we see that

$$q = \perp_{i=1}^{r} \langle a_i\rangle \otimes [1,-w_i]$$

with $w_i = c_i + c_i^2 + d_i^2$ in A, because $2 \times q \sim 0$. Now we can suppose $a_1 = 1$, since $\langle a\rangle \otimes q \cong q$ for all $a \in A^*$. If $r=1$, then $q \sim 0$, i.e. we must have $r \geq 2$. Next we define

$$q_o = [1,-w_1] \perp \langle -1\rangle \otimes [1,-w_2] \perp \dots \perp \langle -1\rangle \otimes [1,-w_r] .$$

This space is isotropic and $a(q) = a(q_o) = 1$. In particular $[q_o] \in IWq(A)_o \cap Wq(A)_t$ and $[q_o]$ is represented by a space of lower dimension than q. Now the proof proceeds by induction, so that we can assume $q_o \sim \langle\langle x,y]]$, and since $2 \times \langle\langle x,y]] \sim 0$, we have $\langle\langle x,y]] \cong \langle\langle a,-w]]$ with $a \in A^*$, $w = c + c^2 + d^2$ (see the proof of (i) ⇒ (ii)). On the other hand it holds

$$q \perp q_o \sim \perp_{i=2}^{r} \langle\langle a_i, -w_i]] ,$$

so that $q \sim \langle\langle a,-w]] \perp \langle\langle a_2,-w_2]] \perp \dots \perp \langle\langle a_r,-w_r]]$. Now we use the following fact: for any $c_1,c_2 \in A^*$ and $u_1,u_2 \in A$ (both of the form $c + c^2 + d^2$) with $1+4u_i \in A^*$ there exists $a' \in A^*$, $w' = c' + c'^2 + d'^2$ (with $1+4w' \in A^*$) such that $\langle\langle c_1,-u_1]] \perp \langle\langle c_2,-u_2]] \sim \langle\langle a',-w']]$. Using this fact and the equivalence above for q we easily deduce by induction that $q \sim \langle\langle e,-v]]$, where $e \in A^*$, $v = c + c^2 + d^2$. This proves (i). Now let us prove the mentioned fact. We have $\langle\langle c_1,-u_1]] \perp \langle\langle c_2,-u_2]] \cong \langle\langle c_1,-u_1]] \perp -\langle\langle c_2,-u_2]] \cong \mathbb{H} \perp q'$, where $\dim q'=6$, $a(q')= 1$ and $w(q') = [(-c_1,u_1] \otimes (-c_2,u_2]] = [(x,y]]$ (here we need our hypothesis (ii)). Hence q' is isotropic (see (4.18), chap.V), and therefore using $I^2Wq(A)_o \cap Wq(A)_t = 0$ we easily deduce, as above, that $q' \sim$

$\langle\langle a', w']]$, where $a' \in A^*$, $w' = c' + c'^2 + d'^2$. This proves the claim.

Let A be a field of characteristic $\neq 2$, such that the classes of quaternion algebras over A form a subgroup of Br(A). Then Elman and Lam have shown that I_A^4 is torsion free and $u(A) \leq 8$. Under the same assumption for a semi local ring A with $2 \in A^*$ the same facts are true, too. But if A is a field of characteristic 2, such that the quaternion algebras over A form a subgroup of Br(A) (actually, are the subgroup $Br(A)_2$), then we do not know, if $I^3Wq(A) = 0$. Probably in this case, it can be $u(A) > 8$. The reason for this assertion is supported by the fact that the proof of $u(A) \leq 8$ in the case $2 \neq 0$ is based on the following result (which immediately follows from (4,21), chap.V): if φ and ψ are two n-folds Pfister spaces over A, $n \geq 2$, then there exists a (n-1)-fold Pfister space τ such that $\varphi \cong \langle\langle a\rangle\rangle \otimes \tau$ and $\psi \cong \langle\langle b\rangle\rangle \otimes \tau$ for some $a, b \in A^*$. If $2 = 0$, then this fact seems not to be true (compare 4,26), chap.V). A thourough treatment of the u-invariant of fields with characteristic $\neq 2$ has been given by Elman and Lam in $[E-L]_1$, $[E-L]_2$.

References

[A] Arf,C.:Untersuchungen über quadratische Formen in Körpern der Charakteristik 2. J.reine und angew.Math.183,148-167 (1941).

[Ar-Pf] Arason,J.K.,Pfister,A.: Beweis des Krullschen Durchschnittssatzes für den Wittring. Inv. Math.12,173-176 (1971).

[A-G] Auslander,M.,Goldman,O.: The Brauer group of a commutative ring. Trans.Am.Math.Soc.97,367-409 (1960).

[B] Baeza,R.: Quadratische Formen über semi lokalen Ringen. Habilitationsschrift, Saarbrücken (1975).

$[B]_1$: Eine Zerlegung der unitären Gruppe über lokalen Ringen. Arch. der Math. Vol.XXIV,144-157 (1973).

$[B]_2$: Eine Bemerkung über quadratische Formen über einem lokalen Ring der Charakteristik 2. Math.Z. 128,363-367 (1972).

$[B]_3$: Über die Torsion der Witt Gruppe Wq(A) eines semi lokalen Ringes. Math.Ann. 207,121-131 (1974).

$[B]_4$: Eine Bemerkung über Pfisterformen. Arch. der Math. Vol.XXV,254-259 (1974).

$[B]_5$: Über die Stufe eines semi lokalen Ringes. Math.Ann. 215,13-21 (1975).

$[B]_6$: Common splitting rings of quaternion algebras over semi local rings. Conference on quadratic forms 1976. Ed. Grace Orzech. Queen's papers in pure and app. math. N°46 (1977).

[B-K] Baeza,R.,Knebusch,M.: Annullatoren von Pfisterformen über semi lokalen Ringen. Math.Z. 140,41-62 (1974).

[Bak] Bak,A. : On modules with quadratic forms. Algebraic K-Theory and its geometric applications. LNM 108, Springer, Berlin - Heidelberg-New York (1969).

[Ba] Bass,H.: Lectures on topics in algebraic K-Theory. Tata Inst. Fund. Res., Bombay (1961).

$[Bo]_1$ Bourbaki,N.: Algèbre commutative. Chap.2. Localisation. Paris Hermann (1961).

$[Bo]_2$: Algèbre. Chap.9. Formes sesquilineaires et formes quadratiques. Paris, Hermann (1959).

[Ca] Cassels,J.W.S.: On the representation of rational functions as sums of squares. Acta Arith.9,79-82 (1964).

[DeM-I] De Meyer,F.,Ingraham,E.: Separable algebras over commutative rings. LNM 181, Springer, Berlin-Heidelberg-New York (1970).

[D]$_1$ Dieudonné,J.: La geométrie des groupes classiques. Springer, Berlin-Heidelberg-New York (1963).

[D]$_2$: Sur les groupes classiques. Publ.Inst.Math.Univ. de Strassbourg. Hermann, Paris (1967).

[E] Eichler,M.: Quadratische Formen und orthogonale Gruppen. Springer, Berlin-Heidelberg-New York (1952).

[E-L]$_1$ Elman,R.,Lam,T.Y.: Quadratic forms and the u-invariant I. Maht. Z. 131,283-304 (1973).

[E-L]$_2$: Quadratic forms and the u-invariant II. Inv.Math.21,125-137 (1973).

[E-L]$_3$: Pfister forms and K-theory of fields. Journal of algebra,23,181-213 (1972).

[E-L]$_4$: Classification theorems for quadratic forms over fields. Comm.Math.Helv. 49,373-381 (1974).

[E-L]$_5$: Quadratic forms over formally real fields and pythagorean fields. Amer.J.of Math.94,1155-1194 (1972).

[E-L]$_6$: Quadratic forms under algebraic extensions. Math.Ann.219,21-42 (1976).

[E-L]$_7$: On the quaternion symbol homomorphism g_F: $k_2F \rightarrow B(F)$. Algebraic K-Theory II. LNM 342. Springer, Berlin-Heidelberg-New York (1973).

[E-L-W] Elman,R.,Lam,T.Y.,Wadsworth,R.: Amenable Fields and Pfister Extensions. Conference on Quadratic Forms. Ed. G. Orzech. Queen's papers in pure and appl.Math. N°46 (1977).

[G] Grothendieck,A.: Le groupe de Brauer I,II,III. Dix exposés sur la cohomologie des schémas. North-Holland,Amsterdam (1968).

[H] Hurwitz,A.: Über die Komposition der quadratischen Formen. Math.Ann.88,1-25 (1923).

[Ka] Kaplanski,I.: Quadratic forms. J.math.soc.Japan 6,200-2o7 (1953).

[K] Knebusch,M.: Bemerkungen zur Theorie der quadratischen Formen über semi-lokalen Ringen. Schriften des math.Inst. der Univ. des Saarlandes, Saarbrücken (1971).

[K]$_1$: Grothendieck- und Wittringe von nicht ausgearteten symmetrischen Bilinearformen. Sitzber.Akad.Wiss.3.Abh., 93-157 (1969/70).

[K]$_2$: Isometrien über semi-lokalen Ringen. Math.Z.108, 255-268 (1969).

[K]$_3$: Runde Formen über semi-lokalen Ringen. Math.Ann. 193,21-34 (1971).

[K]$_4$: Generalisation of a theorem of Artin-Pfister to arbitrary semi local rings and related topics. J.of Algebra, 36,46-67 (1975)

[K]$_5$: Real closures of commutative rings I. J.reine und ang.Math.274/275, 61-89 (1975).

[K]$_6$: Real closures of commutative rings II. J.reine und ang.Math.286/287, 278-313 (1976).

[K]$_7$: Symmetric bilinearforms over algebraic varietes. Conference on quadratic forms 1976. Ed. G.Orzech. Queen's papers in pure and app.Math. N°46 (1977).

[K-R-W]$_1$ Knebusch,M.,Rosenberg,A.,Ware,R.: Signatures on semi local rings. J.of Algebra 26,208-250 (1973).

[K-R-W]$_2$: Structure of Witt Rings and Quotients of abelian Groups Rings. Amer.J.of Math.94,119-155 (1972).

[K-Sch] Knebusch,M.,Scharlau,W.: Über das Verhalten der Witt Gruppe bei Galoischen Körpererweiterungen. Math.Ann.193,189-196 (1971).

[Kne]$_1$ Kneser,M.: Quadratische Formen. Vorlesung SS/WS,1973/74 , Göttingen.

[Kne]$_2$: Witt's Satz für quadratische Formen über lokalen Ringen. Nach.Akad.Wiss.Göttingen,XVII,33-45 (1972).

[Kn] Knus,M.A.: Algébres d'Azumaya et modules projectifs. Comm. Math.Helv.45,372-383 (1970).

[Kn-O] Knus,M.A.,Ojanguren,M.: Théorie de la descente et Algébres d'Azumaya. LNM 383. Springer,Berlin-Heidelberg-New York

[L] Lam,T.Y.: The algebraic theory of quadratic forms. Benjamin (1973).

[Lo] Lorenz,F.: Quadratische Formen über Körpern. LNM 130. Springer Berlin-Heidelberg-New-York (1970).

[Lo-Le] Lorenz,F.,Leicht,J.: Die Primideale des Wittschen Ringes. Inv.Math.10,82-88 (1970).

[Ma] Mandelberg,K.I.: On the classification of quadratic forms over semi local rings. J.of Algebra 33,463-471 (1975).

[M] Milnor,J.: Symmetric inner producs in characteristic 2. Prospects in Math. Annals Study 70, Princeton Univ.Press (1971).

[M-H] Milnor,J.,Husemoller,D.: Symmetric bilinear forms. Springer, Berlin-Heidelberg-New York (1973).

[Mi-V] Micali,A.,Villamayor,O.E.: Algèbres de Clifford et groupe de Brauer. Ann.Scient.Ec.Norm.Sup.4.serie,t.4,285-310 (1971).

[OM] O'Meara,O.T.: Introduction to quadratic forms. Springer, Berlin Heidelberg-New York (1963).

$[Pf]_1$ Pfister,A.: Zur Darstellung von -1 als Summe von Quadraten in einem Körper. J.London Math.Soc.40,159-165 (1965).

$[Pf]_2$: Multiplikative quadratische Formen. Arch.der Math. 16,363-370 (1965).

$[Pf]_3$: Quadratische Formen in beliebigen Körpern. Inv. Math.1,116-132 (1966).

[R-W] Rosenberg,A.Ware,R.: Equivalent topological properties of the space of signatures of a semi local ring. Publicationes Mathematicae 23,283-289 (1977).

[R-Z] Rosenberg,A.,Zelinski,D.: Automorphisms of separable algebras. Pacific Math. Journal 2,1107-1117 (1961).

[Ra] Raynaud,M.: Anneaux locaux henséliens. LNM 169. Springer, Berlin-Heidelberg-New York (1970).

[Ro] Roy,A.: Cancellation of quadratic forms over commutative rings. J.of Algebra 10,116-132 (1968).

[Sa] Sah,C-H.: Symmetric bilinear forms and quadratic forms. J.of Algebra 20,144-160 (1972).

[Sh-W] Shapiro,D.B.,Wadsworth,A.: On multiples of round and Pfister forms. Math.Z.157,53-62 (1977).

[Sch]$_1$ Scharlau,W.: Zur Pfisterschen Theorie der quadratischen Formen. Inv.Math.6,327-328 (1969).

[Sch]$_2$: Quadratic forms. Queen's papers on pure and appl. Math. N°22. Kingston, Canada (1969).

[Se] Serre,J.P.: Cours d'arithmetique. Collection Sup, Presses Univ. de France (1970).

[Sm] Small,Ch.: The group of quadratic extensions. J.pure and appl. Algebra,2,83-105 (1972).

[W] Witt,E.: Theorie der quadratischen Formen in beliebigen Körpern. J.reine und ang.Math.176,31-44 (1937).

[Z-S] Zariski,O.,Samuel,P.: Commutative Algebra I. D.van Nostrand Co. Princeton,New Jersey (1958).

Index

Vol. 489: J. Bair and R. Fourneau, Etude Géométrique des Espaces Vectoriels. Une Introduction. VII, 185 pages. 1975.

Vol. 490: The Geometry of Metric and Linear Spaces. Proceedings 1974. Edited by L. M. Kelly. X, 244 pages. 1975.

Vol. 491: K. A. Broughan, Invariants for Real-Generated Uniform Topological and Algebraic Categories. X, 197 pages. 1975.

Vol. 492: Infinitary Logic: In Memoriam Carol Karp. Edited by D. W. Kueker. VI, 206 pages. 1975.

Vol. 493: F. W. Kamber and P. Tondeur, Foliated Bundles and Characteristic Classes. XIII, 208 pages. 1975.

Vol. 494: A Cornea and G. Licea. Order and Potential Resolvent Families of Kernels. IV, 154 pages. 1975.

Vol. 495: A. Kerber, Representations of Permutation Groups II. V, 175 pages. 1975.

Vol. 496: L. H. Hodgkin and V. P. Snaith, Topics in K-Theory. Two Independent Contributions. III, 294 pages. 1975.

Vol. 497: Analyse Harmonique sur les Groupes de Lie. Proceedings 1973–75. Edité par P. Eymard et al. VI, 710 pages. 1975.

Vol. 498: Model Theory and Algebra. A Memorial Tribute to Abraham Robinson. Edited by D. H. Saracino and V. B. Weispfenning. X, 463 pages. 1975.

Vol. 499: Logic Conference, Kiel 1974. Proceedings. Edited by G. H. Müller, A. Oberschelp, and K. Potthoff. V, 651 pages 1975.

Vol. 500: Proof Theory Symposion, Kiel 1974. Proceedings. Edited by J. Diller and G. H. Müller. VIII, 383 pages. 1975.

Vol. 501: Spline Functions, Karlsruhe 1975. Proceedings. Edited by K. Böhmer, G. Meinardus, and W. Schempp. VI, 421 pages. 1976.

Vol. 502: János Galambos, Representations of Real Numbers by Infinite Series. VI, 146 pages. 1976.

Vol. 503: Applications of Methods of Functional Analysis to Problems in Mechanics. Proceedings 1975. Edited by P. Germain and B. Nayroles. XIX, 531 pages. 1976.

Vol. 504: S. Lang and H. F. Trotter, Frobenius Distributions in GL_2-Extensions. III, 274 pages. 1976.

Vol. 505: Advances in Complex Function Theory. Proceedings 1973/74. Edited by W. E. Kirwan and L. Zalcman. VIII, 203 pages. 1976.

Vol. 506: Numerical Analysis, Dundee 1975. Proceedings. Edited by G. A. Watson. X, 201 pages. 1976.

Vol. 507: M. C. Reed, Abstract Non-Linear Wave Equations. VI, 128 pages. 1976.

Vol. 508: E. Seneta, Regularly Varying Functions. V, 112 pages. 1976.

Vol. 509: D. E. Blair, Contact Manifolds in Riemannian Geometry. VI, 146 pages. 1976.

Vol. 510: V. Poènaru, Singularités C^∞ en Présence de Symétrie. V, 174 pages. 1976.

Vol. 511: Séminaire de Probabilités X. Proceedings 1974/75. Edité par P. A. Meyer. VI, 593 pages. 1976.

Vol. 512: Spaces of Analytic Functions, Kristiansand, Norway 1975. Proceedings. Edited by O. B. Bekken, B. K. Øksendal, and A. Stray. VIII, 204 pages. 1976.

Vol. 513: R. B. Warfield, Jr. Nilpotent Groups. VIII, 115 pages. 1976.

Vol. 514: Séminaire Bourbaki vol. 1974/75. Exposés 453 – 470. IV, 276 pages. 1976.

Vol. 515: Bäcklund Transformations. Nashville, Tennessee 1974. Proceedings. Edited by R. M. Miura. VIII, 295 pages. 1976.

Vol. 516: M. L. Silverstein, Boundary Theory for Symmetric Markov Processes. XVI, 314 pages. 1976.

Vol. 517: S. Glasner, Proximal Flows. VIII, 153 pages. 1976.

Vol. 518: Séminaire de Théorie du Potentiel, Proceedings Paris 1972–1974. Edité par F. Hirsch et G. Mokobodzki. VI, 275 pages. 1976.

Vol. 519: J. Schmets, Espaces de Fonctions Continues. XII, 150 pages. 1976.

Vol. 520: R. H. Farrell, Techniques of Multivariate Calculation. X, 337 pages. 1976.

Vol. 521: G. Cherlin, Model Theoretic Algebra – Selected Topics. IV, 234 pages. 1976.

Vol. 522: C. O. Bloom and N. D. Kazarinoff, Short Wave Radiation Problems in Inhomogeneous Media: Asymptotic Solutions. V. 104 pages. 1976.

Vol. 523: S. A. Albeverio and R. J. Høegh-Krohn, Mathematical Theory of Feynman Path Integrals. IV, 139 pages. 1976.

Vol. 524: Séminaire Pierre Lelong (Analyse) Année 1974/75. Edité par P. Lelong. V, 222 pages. 1976.

Vol. 525: Structural Stability, the Theory of Catastrophes, and Applications in the Sciences. Proceedings 1975. Edited by P. Hilton. VI, 408 pages. 1976.

Vol. 526: Probability in Banach Spaces. Proceedings 1975. Edited by A. Beck. VI, 290 pages. 1976.

Vol. 527: M. Denker, Ch. Grillenberger, and K. Sigmund, Ergodic Theory on Compact Spaces. IV, 360 pages. 1976.

Vol. 528: J. E. Humphreys, Ordinary and Modular Representations of Chevalley Groups. III, 127 pages. 1976.

Vol. 529: J. Grandell, Doubly Stochastic Poisson Processes. X, 234 pages. 1976.

Vol. 530: S. S. Gelbart, Weil's Representation and the Spectrum of the Metaplectic Group. VII, 140 pages. 1976.

Vol. 531: Y.-C. Wong, The Topology of Uniform Convergence on Order-Bounded Sets. VI, 163 pages. 1976.

Vol. 532: Théorie Ergodique. Proceedings 1973/1974. Edité par J.-P. Conze and M. S. Keane. VIII, 227 pages. 1976.

Vol. 533: F. R. Cohen, T. J. Lada, and J. P. May, The Homology of Iterated Loop Spaces. IX, 490 pages. 1976.

Vol. 534: C. Preston, Random Fields. V, 200 pages. 1976.

Vol. 535: Singularités d'Applications Differentiables. Plans-sur-Bex. 1975. Edité par O. Burlet et F. Ronga. V, 253 pages. 1976.

Vol. 536: W. M. Schmidt, Equations over Finite Fields. An Elementary Approach. IX, 267 pages. 1976.

Vol. 537: Set Theory and Hierarchy Theory. Bierutowice, Poland 1975. A Memorial Tribute to Andrzej Mostowski. Edited by W. Marek, M. Srebrny and A. Zarach. XIII, 345 pages. 1976.

Vol. 538: G. Fischer, Complex Analytic Geometry. VII, 201 pages. 1976.

Vol. 539: A. Badrikian, J. F. C. Kingman et J. Kuelbs, Ecole d'Eté de Probabilités de Saint Flour V-1975. Edité par P.-L. Hennequin. IX, 314 pages. 1976.

Vol. 540: Categorical Topology, Proceedings 1975. Edited by E. Binz and H. Herrlich. XV, 719 pages. 1976.

Vol. 541: Measure Theory, Oberwolfach 1975. Proceedings. Edited by A. Bellow and D. Kölzow. XIV, 430 pages. 1976.

Vol. 542: D. A. Edwards and H. M. Hastings, Čech and Steenrod Homotopy Theories with Applications to Geometric Topology. VII, 296 pages. 1976.

Vol. 543: Nonlinear Operators and the Calculus of Variations, Bruxelles 1975. Edited by J. P. Gossez, E. J. Lami Dozo, J. Mawhin, and L. Waelbroeck, VII, 237 pages. 1976.

Vol. 544: Robert P. Langlands, On the Functional Equations Satisfied by Eisenstein Series. VII, 337 pages. 1976.

Vol. 545: Noncommutative Ring Theory. Kent State 1975. Edited by J. H. Cozzens and F. L. Sandomierski. V, 212 pages. 1976.

Vol. 546: K. Mahler, Lectures on Transcendental Numbers. Edited and Completed by B. Diviš and W. J. Le Veque. XXI, 254 pages. 1976.

Vol. 547: A. Mukherjea and N. A. Tserpes, Measures on Topological Semigroups: Convolution Products and Random Walks. V, 197 pages. 1976.

Vol. 548: D. A. Hejhal, The Selberg Trace Formula for PSL (2, $\mathbb{R}$). Volume I. VI, 516 pages. 1976.

Vol. 549: Brauer Groups, Evanston 1975. Proceedings. Edited by D. Zelinsky. V, 187 pages. 1976.

Vol. 550: Proceedings of the Third Japan – USSR Symposium on Probability Theory. Edited by G. Maruyama and J. V. Prokhorov. VI, 722 pages. 1976.

Vol. 551: Algebraic K-Theory, Evanston 1976. Proceedings. Edited by M. R. Stein. XI, 409 pages. 1976.

Vol. 552: C. G. Gibson, K. Wirthmüller, A. A. du Plessis and E. J. N. Looijenga. Topological Stability of Smooth Mappings. V, 155 pages. 1976.

Vol. 553: M. Petrich, Categories of Algebraic Systems. Vector and Projective Spaces, Semigroups, Rings and Lattices. VIII, 217 pages. 1976.

Vol. 554: J. D. H. Smith, Mal'cev Varieties. VIII, 158 pages. 1976.

Vol. 555: M. Ishida, The Genus Fields of Algebraic Number Fields. VII, 116 pages. 1976.

Vol. 556: Approximation Theory. Bonn 1976. Proceedings. Edited by R. Schaback and K. Scherer. VII, 466 pages. 1976.

Vol. 557: W. Iberkleid and T. Petrie, Smooth S^1 Manifolds. III, 163 pages. 1976.

Vol. 558: B. Weisfeiler, On Construction and Identification of Graphs. XIV, 237 pages. 1976.

Vol. 559: J.-P. Caubet, Le Mouvement Brownien Relativiste. IX, 212 pages. 1976.

Vol. 560: Combinatorial Mathematics, IV, Proceedings 1975. Edited by L. R. A. Casse and W. D. Wallis. VII, 249 pages. 1976.

Vol. 561: Function Theoretic Methods for Partial Differential Equations. Darmstadt 1976. Proceedings. Edited by V. E. Meister, N. Weck and W. L. Wendland. XVIII, 520 pages. 1976.

Vol. 562: R. W. Goodman, Nilpotent Lie Groups: Structure and Applications to Analysis. X, 210 pages. 1976.

Vol. 563: Séminaire de Théorie du Potentiel. Paris, No. 2. Proceedings 1975-1976. Edited by F. Hirsch and G. Mokobodzki. VI, 292 pages. 1976.

Vol. 564: Ordinary and Partial Differential Equations, Dundee 1976. Proceedings. Edited by W. N. Everitt and B. D. Sleeman. XVIII, 551 pages. 1976.

Vol. 565: Turbulence and Navier Stokes Equations. Proceedings 1975. Edited by R. Temam. IX, 194 pages. 1976.

Vol. 566: Empirical Distributions and Processes. Oberwolfach 1976. Proceedings. Edited by P. Gaenssler and P. Révész. VII, 146 pages. 1976.

Vol. 567: Séminaire Bourbaki vol. 1975/76. Exposés 471-488. IV, 303 pages. 1977.

Vol. 568: R. E. Gaines and J. L. Mawhin, Coincidence Degree, and Nonlinear Differential Equations. V, 262 pages. 1977.

Vol. 569: Cohomologie Etale SGA 4½. Séminaire de Géométrie Algébrique du Bois-Marie. Edité par P. Deligne. V, 312 pages. 1977.

Vol. 570: Differential Geometrical Methods in Mathematical Physics, Bonn 1975. Proceedings. Edited by K. Bleuler and A. Reetz. VIII, 576 pages. 1977.

Vol. 571: Constructive Theory of Functions of Several Variables, Oberwolfach 1976. Proceedings. Edited by W. Schempp and K. Zeller. VI, 290 pages. 1977

Vol. 572: Sparse Matrix Techniques, Copenhagen 1976. Edited by V. A. Barker. V, 184 pages. 1977.

Vol. 573: Group Theory, Canberra 1975. Proceedings. Edited by R. A. Bryce, J. Cossey and M. F. Newman. VII, 146 pages. 1977.

Vol. 574: J. Moldestad, Computations in Higher Types. IV, 203 pages. 1977.

Vol. 575: K-Theory and Operator Algebras, Athens, Georgia 1975. Edited by B. B. Morrel and I. M. Singer. VI, 191 pages. 1977.

Vol. 576: V. S. Varadarajan, Harmonic Analysis on Real Reductive Groups. VI, 521 pages. 1977.

Vol. 577: J. P. May, E_∞ Ring Spaces and E_∞ Ring Spectra. IV, 268 pages. 1977.

Vol. 578: Séminaire Pierre Lelong (Analyse) Année 1975/76. Edité par P. Lelong. VI, 327 pages. 1977.

Vol. 579: Combinatoire et Représentation du Groupe Symétrique, Strasbourg 1976. Proceedings 1976. Edité par D. Foata. IV, 339 pages. 1977.

Vol. 580: C. Castaing and M. Valadier, Convex Analysis and Measurable Multifunctions. VIII, 278 pages. 1977.

Vol. 581: Séminaire de Probabilités XI, Université de Strasbourg. Proceedings 1975/1976. Edité par C. Dellacherie, P. A. Meyer et M. Weil. VI, 574 pages. 1977.

Vol. 582: J. M. G. Fell, Induced Representations and Banach *-Algebraic Bundles. IV, 349 pages. 1977.

Vol. 583: W. Hirsch, C. C. Pugh and M. Shub, Invariant Manifolds. IV, 149 pages. 1977.

Vol. 584: C. Brezinski, Accélération de la Convergence en Analyse Numérique. IV, 313 pages. 1977.

Vol. 585: T. A. Springer, Invariant Theory. VI, 112 pages. 1977.

Vol. 586: Séminaire d'Algèbre Paul Dubreil, Paris 1975-1976 (29ème Année). Edited by M. P. Malliavin. VI, 188 pages. 1977.

Vol. 587: Non-Commutative Harmonic Analysis. Proceedings 1976. Edited by J. Carmona and M. Vergne. IV, 240 pages. 1977.

Vol. 588: P. Molino, Théorie des G-Structures: Le Problème d'Equivalence. VI, 163 pages. 1977.

Vol. 589: Cohomologie l-adique et Fonctions L. Séminaire de Géométrie Algébrique du Bois-Marie 1965-66, SGA 5. Edité par L. Illusie. XII, 484 pages. 1977.

Vol. 590: H. Matsumoto, Analyse Harmonique dans les Systèmes de Tits Bornologiques de Type Affine. IV, 219 pages. 1977.

Vol. 591: G. A. Anderson, Surgery with Coefficients. VIII, 157 pages. 1977.

Vol. 592: D. Voigt, Induzierte Darstellungen in der Theorie der endlichen, algebraischen Gruppen. V, 413 Seiten. 1977.

Vol. 593: K. Barbey and H. König, Abstract Analytic Function Theory and Hardy Algebras. VIII, 260 pages. 1977.

Vol. 594: Singular Perturbations and Boundary Layer Theory, Lyon 1976. Edited by C. M. Brauner, B. Gay, and J. Mathieu. VIII, 539 pages. 1977.

Vol. 595: W. Hazod, Stetige Faltungshalbgruppen von Wahrscheinlichkeitsmaßen und erzeugende Distributionen. XIII, 157 Seiten. 1977.

Vol. 596: K. Deimling, Ordinary Differential Equations in Banach Spaces. VI, 137 pages. 1977.

Vol. 597: Geometry and Topology, Rio de Janeiro, July 1976. Proceedings. Edited by J. Palis and M. do Carmo. VI, 866 pages. 1977.

Vol. 598: J. Hoffmann-Jørgensen, T. M. Liggett et J. Neveu, Ecole d'Eté de Probabilités de Saint-Flour VI - 1976. Edité par P.-L. Hennequin. XII, 447 pages. 1977.

Vol. 599: Complex Analysis, Kentucky 1976. Proceedings. Edited by J. D. Buckholtz and T. J. Suffridge. X, 159 pages. 1977.

Vol. 600: W. Stoll, Value Distribution on Parabolic Spaces. VIII, 216 pages. 1977.

Vol. 601: Modular Functions of one Variable V, Bonn 1976. Proceedings. Edited by J.-P. Serre and D. B. Zagier. VI, 294 pages. 1977.

Vol. 602: J. P. Brezin, Harmonic Analysis on Compact Solvmanifolds. VIII, 179 pages. 1977.

Vol. 603: B. Moishezon, Complex Surfaces and Connected Sums of Complex Projective Planes. IV, 234 pages. 1977.

Vol. 604: Banach Spaces of Analytic Functions, Kent, Ohio 1976. Proceedings. Edited by J. Baker, C. Cleaver and Joseph Diestel. VI, 141 pages. 1977.

Vol. 605: Sario et al., Classification Theory of Riemannian Manifolds. XX, 498 pages. 1977.

Vol. 606: Mathematical Aspects of Finite Element Methods. Proceedings 1975. Edited by I. Galligani and E. Magenes. VI, 362 pages. 1977.

Vol. 607: M. Métivier, Reelle und Vektorwertige Quasimartingale und die Theorie der Stochastischen Integration. X, 310 Seiten. 1977.

Vol. 608: Bigard et al., Groupes et Anneaux Réticulés. XIV, 334 pages. 1977.